Neuroendocrinological Aspects
of Neurosurgery

Proceedings of the Third Advanced Seminar
in Neurosurgical Research
Venice, April 30–May 1, 1987

Edited by

J. D. Pickard, F. Cohadon, J. Lobo Antunes

Acta Neurochirurgica
Supplementum 47

Springer-Verlag Wien New York

Professor John D. Pickard, Wessex Neurological Centre,
Southampton General Hospital, Shirley, Southampton, U.K.

Professor François Cohadon, Neurochirurgie, Hôpital Pellegrin, Bordeaux, France

Professor João Lobo Antunes, Department of Neurosurgery,
University of Lisbon, Hospital de Santa Maria, Lisbon, Portugal

With 60 Figures

ISSN 0065-1419

ISBN-13: 978-3-7091-9064-7   e -ISBN-13: 978-3-7091-9062-3
DOI: 10.1007/978-3-7091-9062-3

Preface

The Third Advanced Seminar in Neurosurgical Research was held in the Palazzo Pisani Moretta, Venice, Italy from 30th April to 1st May, 1987 and was devoted to "Neuroendocrinological Aspects of Neurosurgery". The general aim of these advanced seminars is to bring together European Neurosurgeons interested or involved in research work, either clinically, experimentally or both, in a given field in order to achieve in-depth informal discussions not possible in the more conventional large congress. In particular, these advanced seminars seek to provide high level teaching by experienced basic scientists, to provide "state of the art" assessment of the subject and to highlight areas of controversy that may be suitable for future research. A special effort is made to identify younger neurosurgeons through the auspices of the European Directory of Neurosurgical Research, who have a particular interest in the subject under discussion, not all of whom will have immediate access to the most advanced, modern technology.

The topic of Neuroendocrinology was chosen because clinically it is an area of controversy that should be amenable to application of new techniques, including molecular biology. There is a tendency on the part of Endocrinologists not to appreciate always that disturbances seen by Neurosurgeons may be different and need special attention. Inevitably there has been a little delay in publication of the manuscripts but these have been updated appropriately. Professor Weindl's lecture on the Circumventricular Organs has recently appeared as a full review in the Journal of Cerebral Blood Flow and Metabolism (1987; 7: 663–672). Neurosurgeons everywhere will look forward to Professor Choux' final detailed analysis of his personal series of endocrine disturbances following craniopharyngioma surgery.

We would like to pay particular tribute to Dr. A. Molendini and Miss Malina Mannarino of the Scientific Public Relations Department of FIDIA Research Laboratories who sponsored and organized this very productive meeting. All the participants enjoyed the fruitful interchanges, made all the more memorable by the atmosphere created by its location in Venice. We hope that this publication will stimulate many neurosurgeons to study their patients with endocrine disturbances and possibly improve their methods of management. We are also very grateful to Mrs. S. Perry for all her care with checking the references and manuscripts. Blackwell Scientific Publications Limited have kindly permitted reproduction of the chapter from "Neuroendocrinology" edited by Stafford L. Lightman and B. J. Everitt (1986: pages 5–31).

J. D. Pickard
F. Cohadon
J. Lobo Antunes

on behalf of the Research Committee of the E.A.N.S.

Contents

Listed in Current Contents

Acta Neurochirurgica, Suppl. 47, 1–15 (1990)
© by Springer-Verlag 1990

Neuroendocrine Anatomy of the Hypothalamus *

B. J. Everitt [1] and **T. Hökfelt** [2]

[1] Department of Anatomy, University of Cambridge, Cambridge, U.K.
[2] Department of Histology, Karolinska Institutet, Stockholm, Sweden

Summary

The hypothalamus is a most complex part of the CNS having rich interconnections with forebrain, limbic and brainstem structures. Its outflow is directed in such a way as to influence the endocrine system (via the neurohypophyseal and adenohypophyseal neurosecretory systems), the autonomic nervous system (via projections to preganglionic cell groups in brainstem and spinal cord) and behavioural responses to physiological and environmental cues via its interaction with limbic and somatomotor systems. The chemical identity of many of its neuronal messengers and those of some of its important afferents, such as the monoaminergic neurons, has opened the way to a form of systematic experimental investigation with chemical tools more powerful than those available to neuroendocrinologists in the past. Much of the information which follows has accrued very rapidly through the use of these methods to reveal the rich complexities of neuroendocrine integration.

The hypothalamus is at the centre of neuroendocrine, autonomic and homeostatic regulation. It controls hormone secretion from the anterior and posterior pituitary and also the organized patterns of autonomic events with which endocrine changes are often closely co-ordinated. The hypothalamus also controls or modifies a number of homeostatic processes as disparate as respiration, body temperature regulation and the ingestion of food and water. The integration of somatomotor, autonomic and endocrine responses which characterize the expression of emotional behaviour, often discussed in relation to the limbic system, are also attributable in large part to the hypothalamus. The hypothalamus is able to exert such profound controls over these functions through its rich connections with many parts of the central nervous system (CNS), no-

tably the brainstem reticular formation and autonomic zones, limbic forebrain (particularly the amygdala) and cerebral cortex. Many of these neuroanatomical relationships have long been recognized, but the recent advances in histochemical and immunocytochemical techniques have revealed remarkable detail of the chemical anatomy of the hypothalamus and, therefore, the ways in which neurohumoral mechanisms governing pituitary secretion are organized.

It is with the organization of the hypothalamus, its afferent and efferent connections, chemical neuroanatomy and relationships to the pituitary that this chapter is concerned.

Development, Boundaries and Blood Supply

Development

The hypothalamus develops from the embryonic diencephalon which is itself one division of the most anterior of the three primary brain vesicles, the prosencephalon (the other division being the telencephalon which becomes the cerebral hemispheres). It first appears as the lowest of three swellings in the lateral wall of the third ventricle and it becomes separated from the other two, the epithalamus and the thalamus, by a longitudinal groove called the hypothalamic sulcus (Fig. 1 a). As the hypothalamus and thalamus develop in the lateral walls of the diencephalon (the thalamus, unlike the hypothalamus, developing in close association with the cortex), the central cavity between them is gradually transformed into the slit-like third ventricle (VIII), whose roof is formed by a single layer of ependyma covered by a capillary-rich mesenchyme. The local invagination of these capillaries, together with the underlying ependyma forms the choroid plexus of the third ventricle.

---

* With kind permission from Blackwell Scientific Publications, Oxford, taken from Lightmann, SL, Everitt, BJ (eds) Neuroendocrinology, pp 5–31.

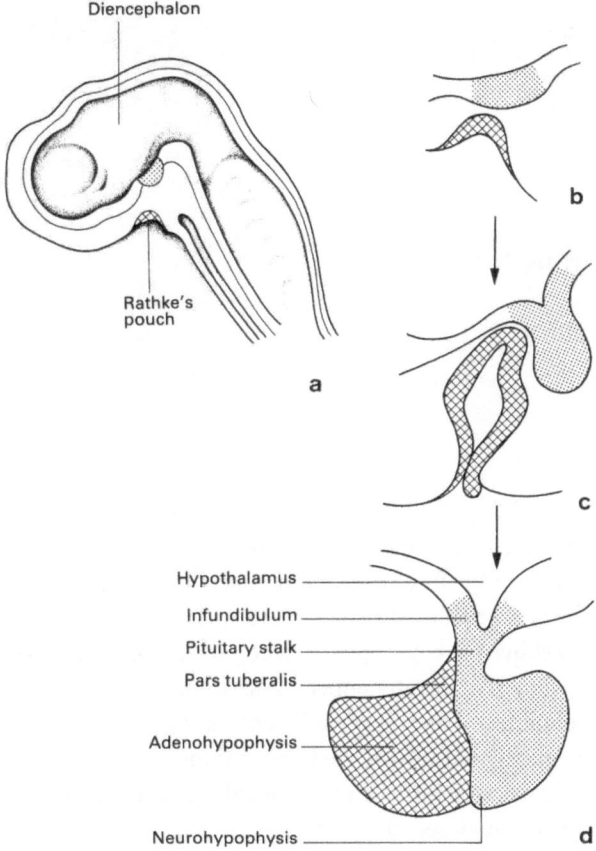

Fig. 1. Development of the pituitary gland. The neurohypophysis is derived from a downward evagination of the diencephalon, the infundibulum (a, b). The adenohypophysis develops from an ectodermal diverticulum of the primitive oral cavity, Rathke's pouch (a, b). The two elements ultimately fuse (c, d) and contact with the oral cavity is lost

The hypophysis develops in close association with the hypothalamus, Indeed, the neurohypophysis (the posterior or neural lobe and infundibular stem) develops as a downward evagination of the diencephalon called the infundibulum (Fig. 1 a). By contrast the adenohypophysis (anterior lobe) has a non-neural origin developing, as it does, from an ectodermal diverticulum of the primitive oral cavity called Rathke's pouch (Fig. 1 b). This diverticulum elongates dorsally to contact the down-growing infundibulum with which it ultimately fuses, losing contact with the oral cavity. Cells from the adenohypophysis subsequently extend along and around the pituitary stalk to form the pars tuberalis (Fig. 1 b). Thus, the neurohypophysis is in neural communication with the overlying hypothalamus, from which it developed as a "downgrowth". The adenohypophysis, conversely, has no direct neural contact with the hypothalamus but instead, a rich vascular

communication between the two develops which is called the portal system. It is by this means that anterior pituitary secretion is controlled by the hypothalamus, that is, they are in humoral contact. (For further embryological details see Heimer 1983, Sidman and Rakic 1982, Hamilton and Mossman 1982.)

Boundaries

From this developmental perspective it can be seen that the hypothalamus is situated at the base of the diencephalon, beneath the thalamus (from which it is separated by the longitudinally running hypothalamic sulcus, which delimits the dorsal extent of the hypothalamus) and on each side of the midline, third ventricle. The rostral boundary of the hypothalamus is conventionally taken to be the lamina terminalis (which appears to run ventrally from the region of the anterior commissure, to the base of the brain) while caudally the boundary is taken to be an imaginary line which extends between the caudal limit of the mamillary body, below, to the posterior commissure above. These rostral and caudal boundaries are not well circumscribed when examining the hypothalamus microscopically, giving the correct impression that it merges in a continuous way with adjacent, surrounding parts of the CNS. Thus, rostrally, the hypothalamus is continuous through the preoptic area (see below) with the septal area and parts of the anterior perforated substance. The anterior zone of the lateral hypothalamus merges laterally with the substantia innominata while more caudally it abuts the internal capsule (dorsolaterally) and globus pallidus (laterally and ventrally). Caudally, the hypothalamus is continuous with the central grey matter and tegmentum of the mesencephalon.

The basal surface of the hypothalamus is defined by the optic chiasm rostrally and the mamillary bodies caudally. In between is a grey swelling called the tuber cinereum which tapers ventrally into the infundibulum which, together with the infundibular part of the adenohypophysis, forms the hypophyseal stalk. This infundibular region at the base of the hypothalamus is frequently called the median eminence. It is a most important site for vascular communication between hypothalamus and pituitary (see below). The lateral eminence found at the lateral margin of the tuber cinereum marks the site of the ventrally located lateral tuberal nuclei of the hypothalamus.

Blood Supply

The blood supply of the basal hypothalamus, infundibulum, hypophyseal stalk and the pituitary gland is

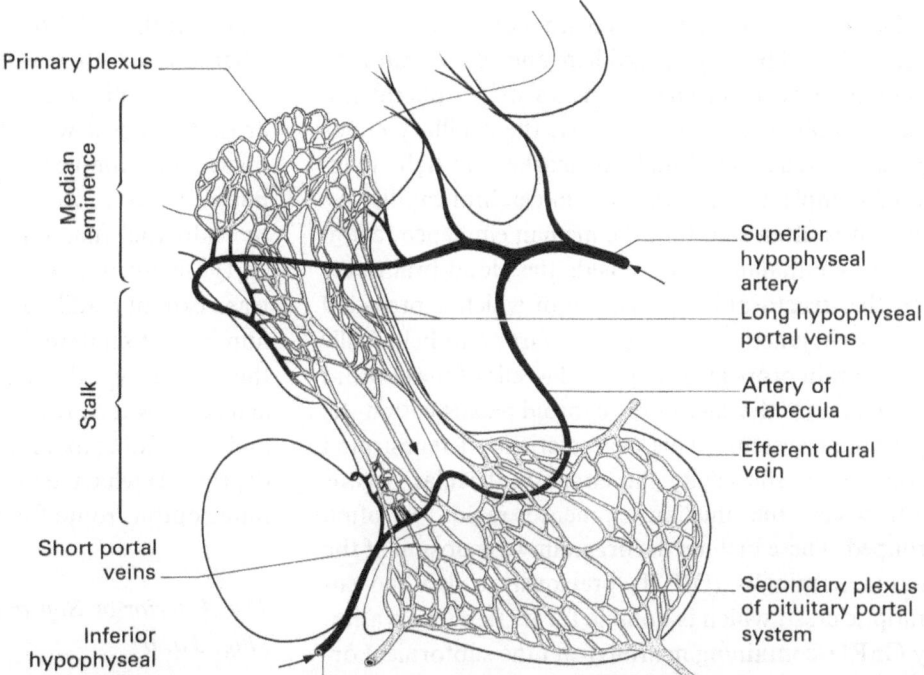

Fig. 2. Schematic representation of the blood supply of the median eminence, infundibulum and adenohypophysis

derived from the carotid arteries via the superior and inferior hypophyseal arteries (Fig. 2). The anterior median eminence and major part of the hypophyseal stalk receive their arterial supply from the superior hypophyseal arteries. The more posterior parts of the median eminence are supplied by separate vessels derived from the circle of Willis, while the most ventral part of the hypophyseal stalk is supplied by the paired trabecular arteries which run from the superior hypophyseal vessels. The neurohypophysis is supplied by the inferior hypophyseal arteries which anastamose with the superior hypophyseal circulation via the trabecular arteries.

The majority of the blood supply of the adenohypophysis is derived not from the hypophyseal arteries directly but indirectly through a capillary portal plexus (Fig. 2). Thus, the arteries supplying the median eminence empty into a rich capillary network. Some of these capillaries form long loops which extend upwards penetrating the palisade zone of the median eminence, almost to the ventricular floor, while others are shorter and remain in a "mantle" plexus. The portal vessels run downwards on both sides of and behind the hypophyseal stalk to reach the adenohypophysis where these so-called "long portal vessels" are joined by the "short portal vessels" arising in the ventral-most hypophyseal stalk and derived in part from the inferior hypophyseal arteries (Fig. 2). Some 90% of adenohypophyseal blood supply arrives through the long portal

vessels (the short portal supply is minor) which distribute blood to the sinusoids of the gland.

It has recently become clear that relatively few venous channels run between the anterior pituitary and cavernous sinus, although this has long been assumed to be the obvious route of venous drainage of the gland, ultimately leaving the skull via the jugular veins. Conversely, numerous venous channels run between the neurohypophysis and cavernous sinus. It has been suggested, on the basis of such observations, that the major adenohypophyseal venous drainage is through the short portal veins into this neurohypophyseal capillary plexus. Thus, the flow of blood containing adenohypophyseal hormones might be regulated in part by those factors normally controlling the neurohypophyseal capillary plexus. Hence, the direction of venous blood flow may vary during different physiological states, indeed might reverse through the capillary plexus to reach the median eminence, or through the arteries that supply large areas of the mediobasal hypothalamus. It has, however, not been established whether such retrograde flow of portal blood occurs in physiological circumstances nor, if it does occur, what significance this may have for the feedback regulation of hypothalamic hormone secretion or for the purported central effects of pituitary peptides. However, there is a clear potential for dynamic, regulatory changes in the pituitary capillary bed which may not previously have been considered in functional studies of the effects of anterior pituitary hormones.

One additional, important feature of the portal capillaries should be emphasized here and that is that they share the structural characteristics of peripheral not central capillaries. That is to say, the capillary endothelium is fenestrated and not marked by tight junctional complexes, as is the case in cerebral capillaries. Thus, it is often said that the median eminence region of the hypothalamus lies outside the blood-brain barrier, the structural manifestation of which is provided by the tight junctions between capillary endothelial cells (in the brain proper) or between the cells of the choroid epithelium (in the case of the choroid plexus within the ventricular system). This characteristic of fenestrated capillaries is shared by the circumventricular organs with which the median eminence region is often grouped. These include the organum vasculosum of the lamina terminalis (OVLT, previously called the supraoptic crest, which is notable for its rich innervation by GnRH-containing neurons) and the subfornical organ (SFO) which is believed to be important in mediating some of the dipsogenic effects of angiotensin. (For further information on the development, gross anatomy and bloody supply of the hypothalamus see Haymaker *et al.* 1969, Moore 1978, Morgane and Panksepp 1979, Krieger and Hughes 1980.)

The ependymal cells at the base of the third ventricle, called tanycytes, are specialized in that they have long processes which extend through the median eminence to reach the portal vessels, terminating in dilated end feet. For many years their function has remained a mystery, but the recent demonstration that they contain the dopamine and adenosine monophosphate regulated phosphoprotein (DARPP-32, see Ouimet *et al.* 1984), which is a specific marker for cells with D1 dopamine receptors, suggests they may mediate some of the effects of dopamine on hypothalamic hormone secretion within the external layer of the median eminence (Everitt *et al.* 1986).

Anatomical Organization

It is convenient to use the three major landmarks on the ventral surface of the hypothalamus (optic chiasm, tuber cinereum and mammillary bodies) to divide the hypothalamus into three parts: rostral *supraoptic,* middle *tuberal* and caudal *mammillary* (Fig. 3). Convention also dictates the longitudinal division of each half of the hypothalamus into medial and lateral zones. The medial zone is cell-rich and the neurons form more or less well-defined aggregations or nuclei. This nuclear organization is rather at the heart of many of the statements in the 1950–60's about hypothalamic function, when this structure was regarded more as a system of "centres". This concept of hypothalamic organization arose because it was observed that lesions of discrete nuclei, for example the ventromedial or medial preoptic nuclei, led to profound changes in specific behaviours or endocrine functions (*e.g.* eating, sexual behaviour and gonadotrophin secretion). The lateral zone, by contrast, cannot readily be subdivided into nuclei, the large numbers of scattered neurons here forming, therefore, the lateral hypothalamic area. This area is also characterized by the presence of a large bundle of ascending and descending axons called the medial forebrain bundle (MFB) which, as will become clear, forms a major input/output route for the medial hypothalamic zone.

The Anterior or Supraoptic Group of Nuclei
(Fig. 3a, b)

Two prominent nuclei are apparent in this group, the supraoptic (SON) and paraventricular (PVN) nuclei. They are prominent because of the large cells which project their axons to the posterior pituitary as the hypothalamo-hypophysial tract. Several islands of magnocellular neurons are also found between the SON and PVN nuclei and appear to be embryonically misplaced parts of the main nuclei. These nuclei are richly vascularized and comprise the magnocellular neurosecretory system. The paraventricular nucleus is larger and more complex than the supraoptic nucleus since it also contains several distinctive parvocellular divisions (anterior, medial, dorsal and lateral—see Swanson and Kuypers 1980, Van den Pol 1982). As will be seen later, the PVN occupies a strategic position in the hypothalamus, having rich interconnections with autonomic and other regions of the brainstem and spinal cord as well as with the pituitary gland (Sawchenko and Swanson 1982). The relationship between the parvocellular and magnocellular divisions of the PVN is not entirely clear, but suggestions that it acts as an "interneuron pool" which integrates information later to be transferred to the magnocellular division can be only partly true (Van den Pol 1980). Immunocytochemical studies have revealed the parvocellular division of the PVN to be a rich source of neuropeptides involved in the regulation of anterior pituitary secretion (see below).

The suprachiasmatic nuclei are also distinctive, almost circular nuclei (in coronal section) which nearly

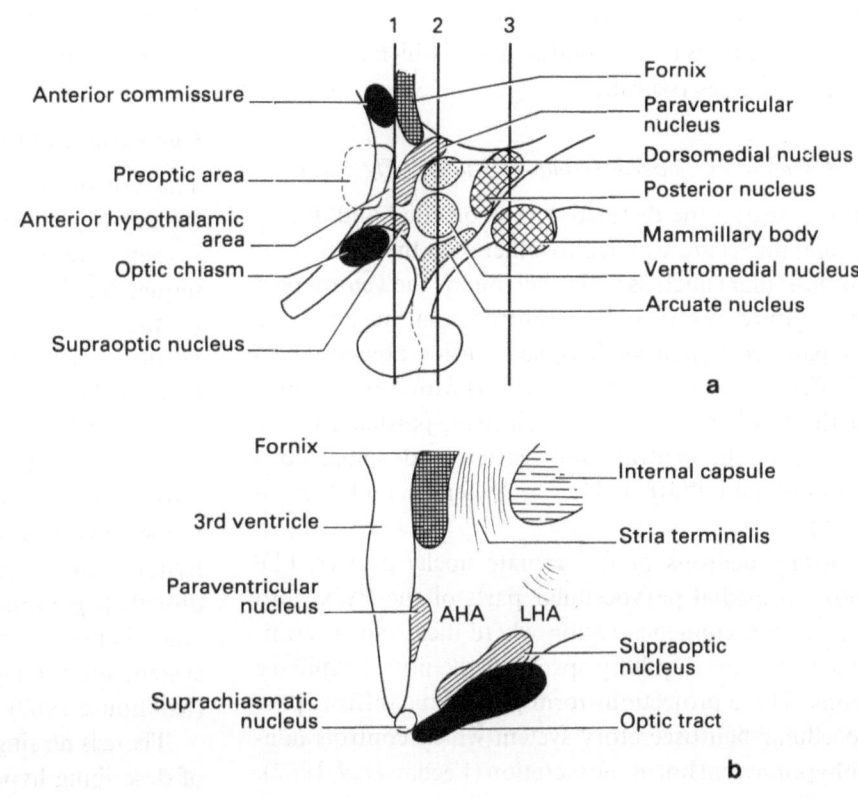

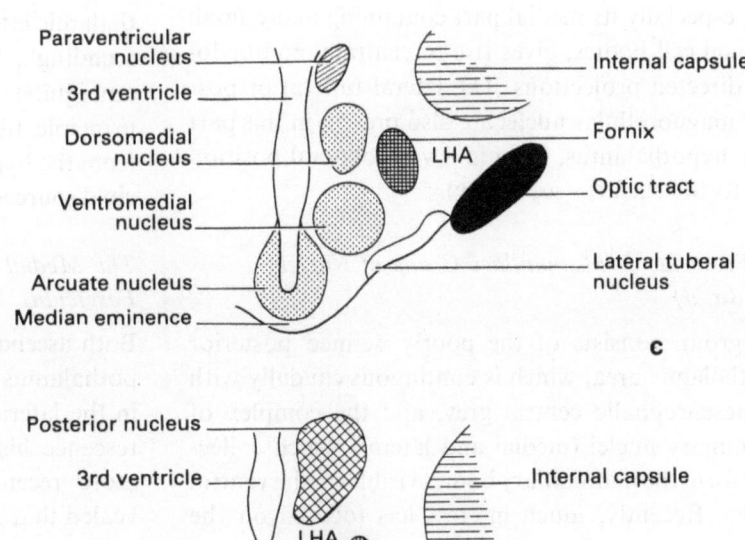

Fig. 3. Schematic representation of hypo-thalamic nuclei. (a) Represents a sagittal view of the hypothalamus while (b), (c), and (d) are coronal sections taken in the planes *1, 2,* and *3* in (a)

touch each other in the midline at the base of the hypothalamus, beneath and adjacent to the rostral part of the third ventricle lying just above the optic chiasm. These nuclei, are intimately concerned with the generation of circadian rhythms and control daily rhythms in hormone secretion and behaviour. Much has been

learned recently of their chemical neuroanatomy (Card and Moore 1984, Lightmann and Everitt 1986).

The continuum formed by the anterior hypothalamic and preoptic areas, including the medial preoptic nucleus, is also an important, if not well defined region in this anterior zone. It represents a major target for the sex steroids, being rich in oestradiol and testosterone receptors, as is the septal area with which the preoptic area merges rostrally.

The Middle or Tuberal Group of Nuclei (Fig. 3a, c)

In this region the distinctive ventromedial and dorsomedial nuclei are situated together with the arcuate (or infundibular) nucleus which surrounds the ventral part of the third ventricle. The ventromedial nucleus, like the paraventricular nucleus, has considerable numbers of afferent and efferent connections with other regions of the CNS including the brainstem, particularly the mesencephalic central grey matter, and spinal cord (Conrad and Pfaff 1976a, b, Swanson and Cowan 1975).

Many neurons of the arcuate nuclei project, like those in medial parvocellular parts of the PVN, into the median eminence region where their axons terminate in the pericapillary spaces of the portal capillary loops. These projections form part of the diffuse, parvocellular neurosecretory system which controls adenohypophyseal hormone secretion (Lechan et al. 1982). It is, however, important to note that the arcuate nucleus, especially its medial part containing many small neuronal cell bodies, gives rise to centrally and/or locally directed projections. The lateral tuberal or posterior magnocellular nuclei are also present in this part of the hypothalamus, lying in a ventrolateral position close to the surface (see below).

The Posterior or Mammillary Group of Nuclei (Fig. 3a, d)

This group consists of the poorly defined posterior hypothalamic area, which is continuous caudally with the mesencephalic central grey, and the complex of mammillary nuclei (medial and lateral) which collectively form the mammillary bodies visible on the ventral surface. Recently, much interest has focused on the posterior magnocellular nuclei since they appear to be comprised, in large part, of neurons containing histamine, GABA and the neuropeptide galanin and which project widely over the forebrain, particularly innervating the neocortex (Takeda et al. 1984).

Throughout the antero-posterior extent of the hypothalamus there lies a sub-ependymal, periventricular nucleus which is rich in neurons, often of characteristic neurochemical type (see below) and which also contains an extensive system of ascending and descending fibres (see below).

(For more details of hypothalamic anatomy see

Haymaker et al. 1969, Reichlin et al. 1978, Moore 1978, Morgane and Panksepp 1979.)

Connections of the Hypothalamus

The use of silver impregnation, histochemical tract-tracing (e.g. horseradish peroxidase transport), fluorescent dye-tracing and immunocytochemical techniques has both confirmed, and revealed new details of, the complexity and richness of hypothalamic connections. These connections are with the forebrain, particularly limbic areas, brainstem and spinal cord structures as well as with the vascular system. The latter are both efferent (e.g. hypothalamic hormones secreted into the portal plexus) and afferent (e.g. steroid hormones involved in feedback actions) in nature. Some hypothalamic afferents enter the medial, nuclear areas directly (e.g. monoaminergic neurons) but many others enter indirectly through the medial forebrain bundle system after relay in the lateral hypothalamic area (Millhouse 1969).

There is no single, totally acceptable or efficient way of describing hypothalamic connections. Many, if not all, of the so-called afferent systems also carry hypothalamic efferents. Terming them "ascending" or "descending", "extrinsic" or "intrinsic" also introduces ambiguities and a list of exceptions. Here, then, the principle fibre systems carrying information to and from the hypothalamus will be described with the principal sources and destinations of their component parts.

The Medial Forebrain Bundle and Dorsal Longitudinal Fasciculus

Both ascending and descending projections to the hypothalamus run in the MFB and, in many cases, relay in the lateral hypothalamic area. The advent of fluorescence histochemical techniques in the 1960s and, more recently, immunocytochemical techniques revealed that a large proportion of the monoamine-containing neurons in the brainstem (Fig. 4) project directly to the hypothalamus, as well as the striatum, basal forebrain and neocortex, through the MFB (Fig. 5). Noradrenergic neurons (Fig. 4a) in the ventrolateral medulla oblongata (group A1 in the nomenclature of Dahlstrom and Fuxe 1964 and which will be referred to in the rest of the description which follows; see Hökfelt et al. 1984b) and in the nucleus of the solitary tract (group A2) project heavily to the hypothalamus. The axons ascend in the central tegmental tract to enter the MFB in the midbrain (Moore and Bloom 1978). This group of fibres is often termed the ventral noradrenergic bundle/pathway to distinguish it from the projections of the locus coeruleus (cell group A6, Fig. 4b), which run in the so-called dorsal noradrenergic bundle. However, it has recently become clear that a substantial number of A1 axons run up to join the dorsal bundle in the mesencephalon, subsequently merging with the MFB in the caudal, lateral hypothalamus (Swanson and Sawchenko 1983). While the locus coeruleus noradrenergic neurons densely and uniquely innervate the neocortex and hippocampus, they also project substantially to the periventricular zone of the hypothal-

amus (Moore and Bloom 1979, Sawchenko and Swanson 1982).

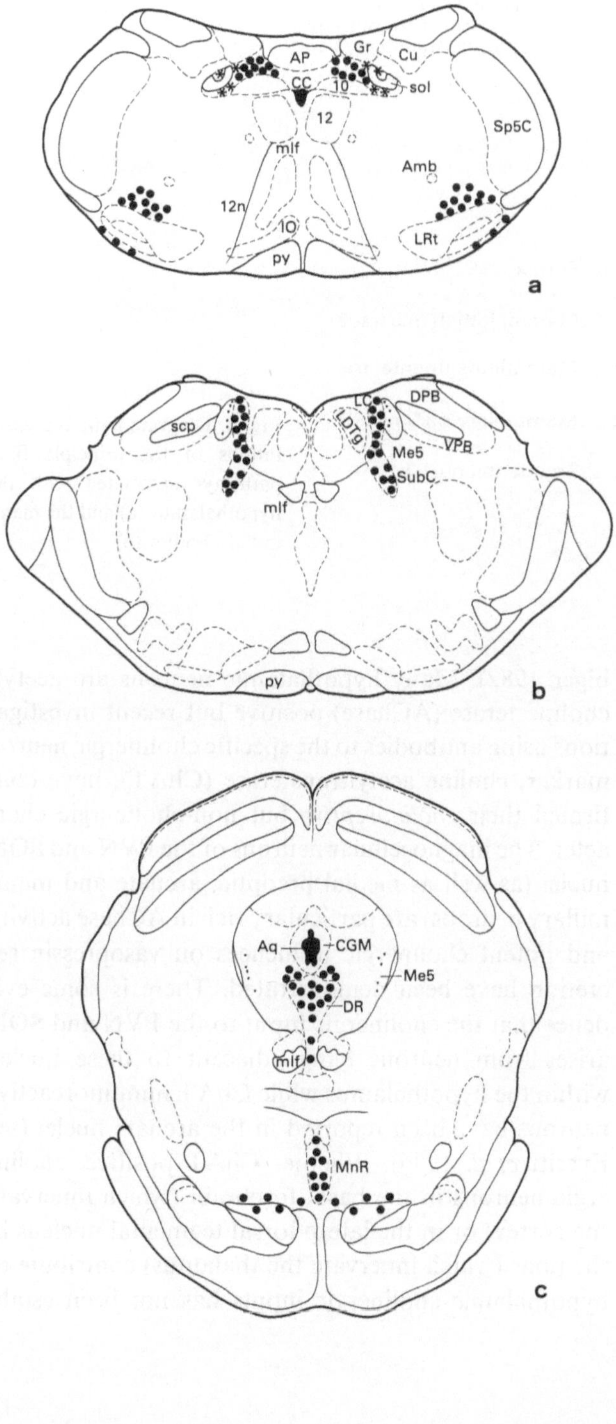

Fibres leave the MFB to pass dorsally and medially over the fornix to reach more dorsal areas of the hypothalamus (e.g. the paraventricular nucleus) or sweep ventrally and medially to reach more ventral areas (e.g. supraoptic, arcuate and suprachiasmatic nuclei). The medial preoptic and anterior hypothalamic areas also receive a rich noradrenergic input. The particularly dense noradrenergic innervation of the supraoptic and paraventricular nuclei (both parvi- and magnocellular parts of the latter) has been analysed in great detail (Sawchenko and Swanson 1982).

The adrenaline-containing neurons of the medulla (cell groups C_1 and C_2; Hökfelt et al. 1984a; Fig. 4a), which form rostral extensions of groups A_1 and A_2, also project to the medial hypothalamus, again with dense terminations in the paraventricular nuclei. This system of projections is often neglected in functional terms because it is assumed to be small. However, optimizing the immunohistochemical visualization of the marker, biosynthetic enzyme phenylethanolamine-n-methyl transferase (PNMT) has revealed that both the neuronal cell groups and their projections are substantial and focused largely on the hypothalamus (Hökfelt et al. 1984a).

The 5-hydroxytryptamine (5-HT)-containing neurons of the mesencephallic raphe nuclei (cell groups B7, 8 and 9; Dahlstrom and Fuxe 1964; Fig. 4c) also project to the hypothalamus via the MFB. There is a widespread innervation of the medial hypothalamus, but this is particularly dense in the mammillary complex, the periventricular zone, the arcuate and suprachiasmatic nuclei. The internal layer of the median

Fig. 4. Three transverse sections of the brainstem through the medulla oblongata (a), the pons (b) and the midbrain (c). In (a), catecholamine cell groups A1 (and C1) in the ventrolateral medulla and A2 (and C2) in the nucleus tractus solitarius are shown. In (b), the noradrenergic locus ceruleus (A6) is shown in the floor of the fourth ventricle. In (c), the serotoninergic cell groups B7 (dorsal raphe nucleus) and B8 (median raphe nucleus) are shown. ● Represents bodies; *Amb* nucleus ambiguus; *AP* area postrema; *Aq* cerebral aqueduct; *CC* central canal; *CGM* periaqueductal grey matter; *Cu* cuneate nucleus; *DR* dorsal raphe nucleus; *Gr* gracile nucleus; *IO* inferior olive; *LC* locus ceruleus; *LRt* lateral reticular nucleus; *LDTg* latero-dorsal tegmental nucleus; *mlf* medial longitudinal fasciculus; *MnR* median raphe nucleus; *Me5* mesencephalic nucleus of fifth nerve; *DPB*, *VPB* dorsal and ventral parabrachial nuclei; *sol* solitary tract; *Sp5C* spinal nucleus of fifth nerve; *SubC* nucleus subceruleus; *scp* superior cerebellar peduncle; *py* pyramid; *10, 12* vagal and hypoglossal nuclei

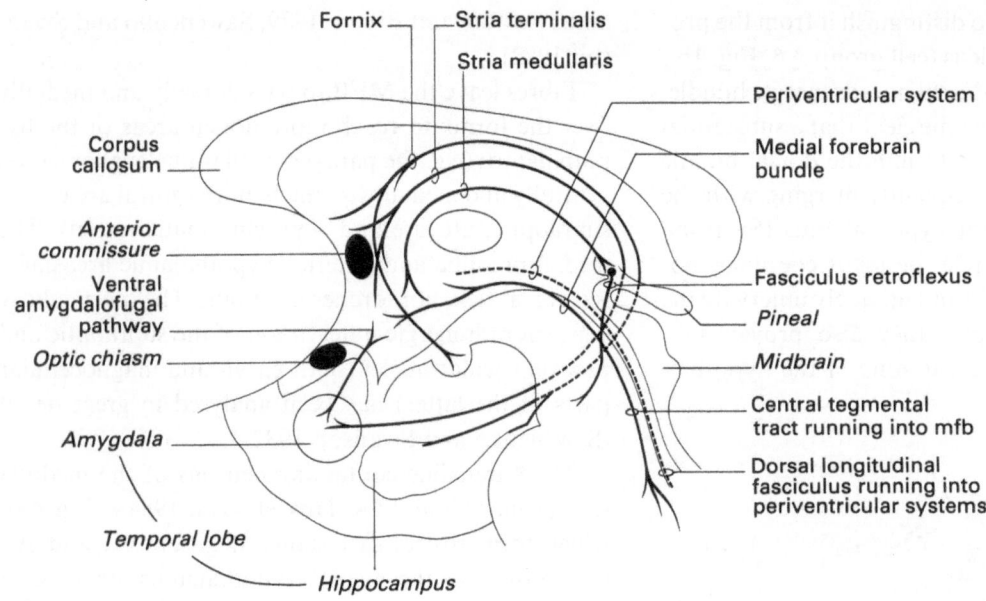

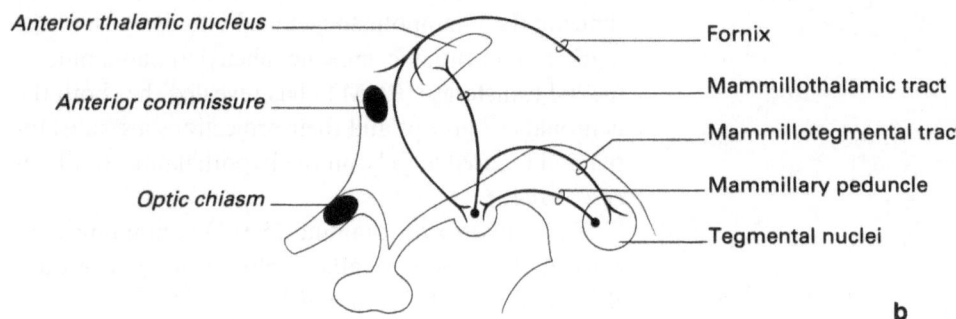

Fig. 5. Diagrammatic representations of the principle fibre pathways associated with the hypothalamus (a) and the mammillary bodies (b)

eminence also receives a moderate 5-HT input (Steinbusch 1981).

The dorsal longitudinal fasciculus is part of an extensive periventricular system of ascending and descending fibres connecting the hypothalamus with the mesencephalic periaqueductal grey and preganglionic, autonomic nuclei in the brainstem and spinal cord. Many noradrenergic (A_1, A_2, and A_6), serotoninergic and also adrenergic axons ascend from the brainstem to reach the periventricular nuclear zone of the hypothalamus through the system. Many dopamine-containing neurons lie within the periaqueductal and periventricular (hypothalamic) grey and these, too, project to and within the hypothalamus, particularly to its rostral parts, although detailed information concerning this system is generally lacking (Lindvall and Björklund 1978, Moore and Bloom 1978).

The source of cholinergic afferents to the hypothalamus has yet to be definitively established (see Fi-

biger 1982). Many hypothalamic neurons are acetylcholinesterase (AChase) positive but recent investigations using antibodies to the specific cholinergic neuron marker, choline acetyltransferase (ChAT), have confirmed their *cholinoceptive* but non-cholinergic character. The magnocellular neurons of the PVN and SON nuclei (as well as medial preoptic, arcuate and mammillary neurons) are particulary rich in AChase activity and potent cholinergic influences on vasopressin secretion have been demonstrated. There is some evidence that the cholinergic input to the PVN and SON arises from neurons lying adjacent to these nuclei, within the hypothalamus while ChAT immunoreactive neurons have been reported in the arcuate nuclei (see Everitt *et al.* 1986). Whether ChAT positive, cholinergic neurons in the basal forebrain (which innervate the cortex) or in the laterodorsal tegmental nucleus in the pons (which innervate the thalamus) contribute to hypothalamic cholinergic inputs has not been established. Powerful influences of acetylcholine on neurohypophyseal, adenohypophyseal and behavioural regulation make this an important area to clarify. Great advances in revealing the details of cholinergic neuronal systems have followed the development of specific ChAT antibodies and it seems reasonable to suggest that the source of hypothalamic cholinergic inputs will become clear relatively soon (see Mesulam *et al.* 1984).

Many non-monoaminergic brainstem neurons, often described as "reticular formation neurons" project via the central tegmental tract and MFB. A large proportion terminate on lateral hypothalamic neurons, or the dendrites of medial hypothalamic neurons which extend into the lateral hypothalamic area. Lateral hypothalamic neurons have widespread projections, some of which are directed medially to the nuclear zone of the hypothalamus (Millhouse 1969, Moore 1978).

Of the descending inputs to the hypothalamus which run in the MFB, the most important arise from basal forebrain structures, often grouped together under the confusing heading "limbic system". These include the olfactory areas of the forebrain, the amygdaloid nuclear complex, septal nuclei and, indeed, neocortical structures. While some of these structures, *e.g.* the amygdala (see below) project directly into the medial hypothalamus, many terminate in the lateral hypothalamic area to reach more medial structures indirectly.

Of course, many axons running in the MFB may not be involved with hypothalamic function at all. These include descending projections from the forebrain which run to the brainstem reticular formation and autonomic nuclei and also ascending systems such as the nigrostriatal dopaminergic system and monoaminergic projections to the cortex.

The lateral hypothalamic area and MFB system also provide an important route for the flow of information out of the medial hypothalamus. Although some of the medial nuclei project directly to structures outside the hypothalamus (see below), the lateral hypothalamic projection forms a major pathway for medial hypothalamic influences on other areas of the CNS, including autonomic nuclei in the brainstem and spinal cord. However, some structures, such as the parvocellular division of the paraventricular nucleus, also project directly to areas including the dorsal vagal complex and intermediolateral cell column in the thoracic and lumbosacral spinal cord which are intimately concerned with autonomic function.

Stria Terminalis and Ventral Amygdalofugal Pathway

The stria terminalis is a major pathway reciprocally interconnecting the amygdaloid nuclear complex in the temporal lobe with the medial hypothalamus (Fig. 5a). The organization of the amygdala is too complex to describe in detail here but, to simplify it in the present context, it is the corticomedial and central nuclei which project in major part to the hypothalamus. The stria terminalis divides into several components in the region of the anterior commissure (having passed its looping course from the temporal lobe, behind and then above the thalamus). Some of its components pass to the septal area, the bed nucleus of the stria terminalis and the ventral striatum while another runs downwards and caudally into the medial hypothalamus (Fig. 5a). The amygdala, including basolateral nuclei, also projects to the hypothalamus via a diffuse system called the ventral amygdalofugal pathway which runs between the medial temporal lobe and lateral hypothalamic area/MFB system. Fibres running in this system, besides terminating in the lateral and medial hypothalamus, also run to the mediodorsal nucleus of the thalamus, ventral striatum, cingulate cortex and brainstem autonomic nuclei. Clearly, the amygdala directs a considerable proportion of its efferents to the hypothalamus where it may modulate both anterior pituitary hormone secretion and emotional behaviour. A major source of afferents to those parts of the amygdala projecting to the hypothalamus are the olfactory system (to the cortical and medial nuclei) and gustatory parts of the autonomic nervous system (nucleus of the solitary tract and parabrachial nuclei which project to the central nucleus). (For further details of amygdalo-hypothalamic connections see Heimer 1983.)

Fornix and Connections of the Mammillary Nuclei

The fornix is the major efferent projection from the hippocampus in the temporal lobe. It runs behind and then above the thalamus to divide over the anterior commissure. The pre-commissural division terminates in septal nuclei and ventral striatum. The post commissural division terminates within the mammillary nuclei (Fig. 5a, b).

The mammillary peduncle contains axons of neurons in the mesencephalic tegmental nuclei which terminate primarily in the lateral mammillary nucleus (Fig. 5b). The majority of efferents from the mammillary complex run in the mammillothalamic tract to reach, via the internal medullary lamina, the anterior thalamic nucleus (Fig. 5b). (The multisynaptic projections of the latter via the cingulate cortex, cingulum and entorhinal cortex which projects in turn to the hippocampal formation, constitutes the long known and still poorly understood "Papez circuit".) The mam-

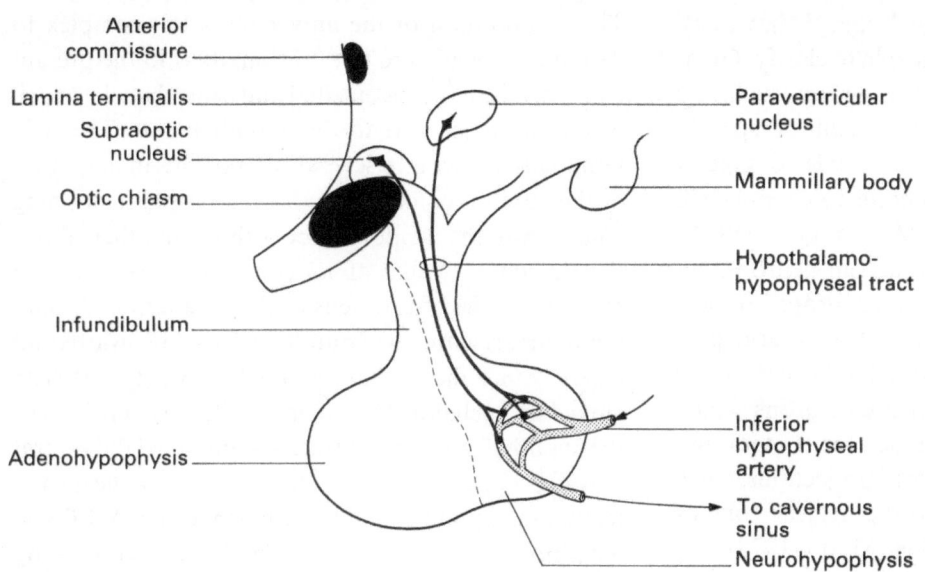

Fig. 6. Schematic representation of the magnocellular neurosecretory system. Neurons in the supraoptic and paraventricular nuclei send their axons via the hypothalamo-hypophyseal tract through the infundibulum to the neurohypophysis where terminals lie in association with capillary walls, the site of neurosecretion

millotegmental tract (Fig. 5b) is largely composed of collaterals of mammillothalamic fibres which project to the midbrain tegmental nuclei where the mammillary peduncle arises (Heimer 1983).

Stria medullaris

While often described as a projection system from the septal nuclei and medial olfactory stria to the habenula, many of the component fibres of the stria medullaris also connect the lateral preoptic-anterior hypothalamic region with the epithalamic, habenular nuclear complex (Fig. 5a). The habenula projects to the interpeduncular nucleus and surrounding ventral tegmental area of the midbrain, that is to say, into an area with important projections back into the circuitry of the limbic system.

Retinohypothalamic Tract

The suprachiasmatic nuclei receive a direct, bilateral input from the retina. This projection is most important in mediating the influence of day/night cycles on the entrainment of circadian rhythms.

Intrahypothalamic Connections

There are abundant intrahypothalamic projections which are both too numerous, complex and controversial to detail (see Conrad and Pfaff 1976a, b, Swanson 1976, Saper, Swanson and Cowan 1976, Swanson and Cowan 1975) although a few will be highlighted here. The medial preoptic and anterior hypothalamic areas, which are rich in steroid hormone receptors and implicated in the control of sexual behaviour, have substantial projections into the lateral hypothalamic area along its entire axis. In addition, the medial preoptic area projects caudally to the anterior hypothalamic area, paraventricular nucleus, arcuate nucleus and both internal and external layers of the median eminence (see below). The anterior hypothalamic area projects to the preoptic region rostrally, and also to the paraventricular, dorsomedial, ventromedial and arcuate nuclei. The suprachiasmatic nuclei project to the paraventricular nuclei. The ventromedial nucleus has received much attention in recent years and, apart from ascending projections to the preoptic/anterior hypothalamic and lateral hypothalamic areas, gives important inputs to the mesencephalic periaqueductal grey which is continuous with the periventricular, posterior hypothalamus.

Chemical Neuroanatomy and the Hypothalamohypophyseal Pathways

The Magnocellular Neurosecretory System

This is comprised of the projections of the large neurons of the paraventricular and supraoptic nuclei (and associated cell clusters in the anterior medial hypothalamus) to the neurohypophysis (see Sofroniew 1985). Axons of paraventricular neurons sweep laterally, over and around the fornix and then ventrally to join axons of the supraoptic neurons as the supraoptico- (or hypothalamo-) hypophyseal tract (Fig. 6). The axons then turn caudally, to run in the internal layer of the median eminence to reach the neurohypophysis. Here, the terminals assume a close association with the capillary plexus of the posterior lobe.

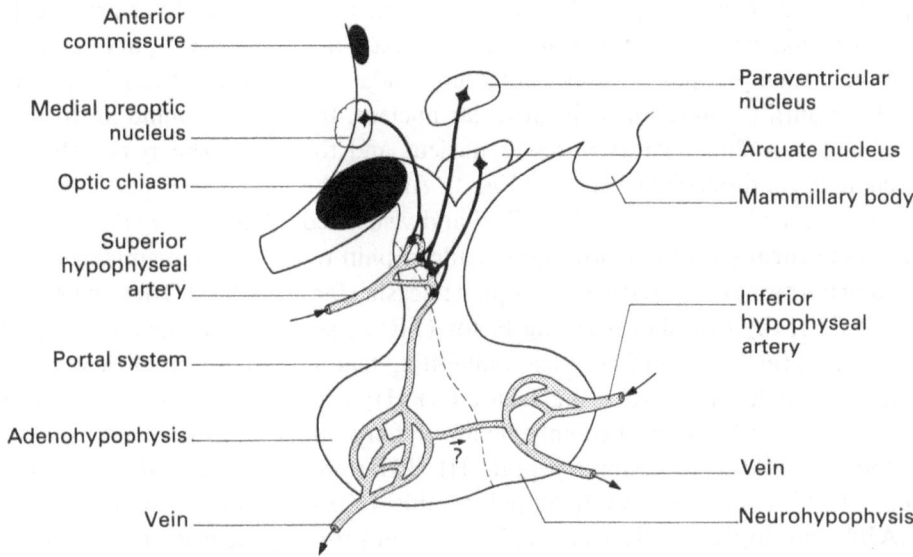

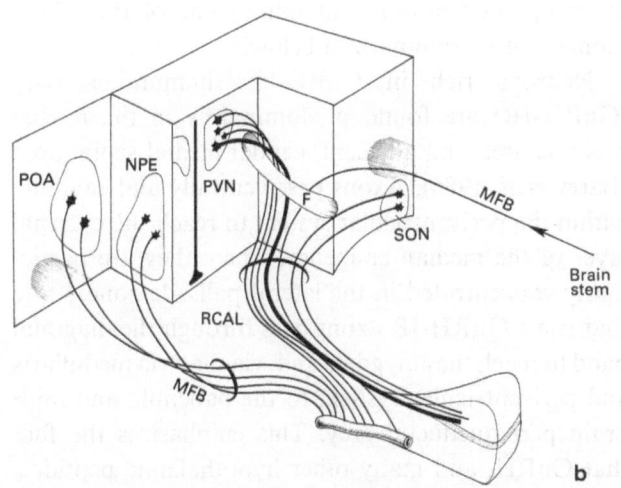

Fig. 7. (a) Schematic representation of the parvocellular neurosecretory system. Neurons in the medial preoptic area (*e.g.* GnRH-containing), anterior periventricular nucleus (*e.g.* somatostatin-containing), medial parvocellular paraventricular nucleus (*e.g.* CRF-, TRF- and neurotensin-containing) and arcuate nucleus (*e.g.* GRH-, GABA- and dopamine-containing) send their axons down to the portal vessels in the external layer (palisade zone) of the median eminence where neurosecretion occurs. (b) Further illustration of both parvocellular and magnocellular neurosecretory systems showing the course frequently taken by neurosecretory neurons over and around the fornix, converging in the retrochiasmatic area eventually to reach the external and internal layers of the median eminence and infundibulum (after Palkovits 1982). *RCAL* retrochiasmatic area; *NPE* periventricular nucleus. Other abbreviations in text

Approximately equal numbers of oxytocin- and vasopressin-containing neurons are present in both supraoptic and paraventricular nuclei of the rat, albeit in separable zones (Rhodes *et al.* 1982, Sawchenko and Swanson 1982). Thus, oxytocin neurons are present in more dorsal parts of the supraoptic nucleus, they predominate in anterior and medial magnocellular parts of the paraventricular nucleus and group together in ventromedial parts of its posterior magnocellular division (Sawchenko and Swanson 1982). Vasopressin neurons are largely restricted to ventral parts of the supraoptic nucleus and a spherical, lateral zone in the posterior magnocellular paraventricular nucleus (Sawchenko and Swanson 1982, Sofroniew 1985). The appropriate neurophysins are also present, and readily visualized in these oxytocin and vasopressin neuronal cell bodies, axons and terminals.

The Parvocellular Neurosecretory System

This is less concisely organized if, indeed, it is an entity and has only recently been revealed in all its complexity by contemporary immunocytochemical techniques which have allowed the visualization of hypothalamic peptides and other humoral substances directly. In essence, the term describes that disparate system of neurons (in the hypophysiotrophic areas—Szentagothai *et al.* 1972) projecting to the external layer of the median eminence where their terminals release their peptidergic, aminergic and, probably, amino acid contents into the portal vessels for carriage to the adenohypophysis (Fig. 7a). It has been pointed out (Palkovits 1982) that many of these systems converge on the retrochiasmatic area during their projection to the median eminence and neurohypophysis (Fig. 7b). Subsequently, four medial hypothalamic nuclear structures have become the focus of attention in this context. They are the medial preoptic area (and adjacent, caudal medial septum), the anterior periventricular nucleus, the medial parvocellular paraventricular nucleus and the arcuate (infundibular) nucleus (see Fig. 3). Among the hypothalamic hormones now identified and visualized in these neurons are the following: gonadotrophin releasing hormone (GnRH); corticotrophin releasing factor (CRF); thyrotrophin releasing factor (TRH); somatostatin (growth hormone release-inhibiting factor, SS); growth hormone releasing factor (GRH); neurotensin (NT); Met- and Leu-enkephalin; dynorphin; β-endorphin/ACTH; galanin; peptide HI (PHI) and the related vasoactive intestinal polypeptide (VIP); GABA and dopamine. This incomplete list will probably enlarge considerably in the future as new biologically active peptides are discovered. The localization of groups of neurons containing some of these hormones will be summarized below.

Neurons rich in GnRH-like immunoreactivity (GnRH-IR) are found predominantly in the medial preoptic area and adjacent, caudal medial septal area (Barry et al. 1985). Axons pass ventrally and caudally within the periventricular system to reach the external layer of the median eminence, where they are particularly concentrated in the lateral palisade zone. Note that some GnRH-IR axons pass through the diagonal band to reach the amygdala and, via the stria medullaris and periventricular system, to the habenula and midbrain periaqueductal grey. This emphasizes the fact that GnRH, and many other hypothalamic peptides, probably serve a neurotransmitter/neuromodulator role in addition to the humoral functions normally associated with adenohypophyseal regulation. Suggestions of a group of GnRH-IR neurons in the arcuate nucleus remain contentious, at least in the rat, although in the primate its status is somewhat more secure.

Numerous neurons containing somatostatin-like immunoreactivity (SS-IR) are present throughout the anterior periventricular nucleus, extending back to the periventricular zone of the paraventricular nucleus (see Johansson et al. 1984). There are also large numbers of SS-IR neurons in the arcuate nucleus. Selective lesioning and retrograde transport studies have revealed, however, that the majority of SS-IR neurons projecting to the external layer of the median eminence, where an extremely rich terminal plexus exists throughout its mediolateral and antero-posterior extent, lie within the anterior, periventricular nucleus. The axons course first laterally, over the fornix, to dip ventrally, medially and

caudally to arrive at the median eminence—a trajectory reminiscent of magnocellular paraventricular neuronal projections (Fig. 7b). Many small, SS-IR neurons are also found in the suprachiasmatic nucleus.

The parvocellular division of the paraventricular nucleus, particularly the medial subnucleus, has proved to be very rich in peptidergic neurons projecting to the median eminence portal vessels (e.g. Sawchenko et al. 1984). Thus, major groups of CRF-, CCK-, TRH-, NT-, enkephalin (ENK)-, PH1 and galanin immunoreactive neurons all lie within its boundaries and have abundant terminals around the portal vessels of the palisade zone of the external layer of the median eminence. The course taken by axons en route to this site appears, at least in some cases, to resemble that for somatostatin neurons (see above and Fig. 7b). In addition, parvocellular vasopressin (AVP-IR) neurons are found within the paraventricular nucleus and some of these, too, project to the external layer of the median eminence rather than to the neurohypophysis (Sawchenko et al. 1984).

The arcuate (infundibular) nucleus has long been the focus of attention for neuroendocrinologists. Many of its neurons have been demonstrated to project to the external layer of the median eminence—so called tuberoinfundibular or tuberohypophyseal neurons (Lechan et al. 1982). Of these, the tuberoinfundibular dopamine neurons (TIDA neurons, cell group A12) have attracted much interest since dopamine released from this site into the portal vessels has proved to be a most, if not the most, important prolactin inhibitory factor. Some of these neurons are oestradiol target neurons and they also contain additional substances with potential actions on the adenohypophysis (see below). Neurons containing NT, GABA, GRH, galanin, Met-enkephalin, dynorphin, β-endorphin, somatostatin and other peptides (Everitt et al. 1986) are all found within the boundaries of this small nucleus, and many of these compounds are also within terminals in the external layer of the median eminence.

The major group of β-endorphin-containing neurons in the CNS is also localized within the arcuate nucleus and in an area lateral to it. These neurons project widely over the neuraxis, both to hypothalamic nuclei (paraventricular, medial preoptic and locally within the arcuate nuclei) and elsewhere (amygdala, periaqueductal grey). There is some controversy as to whether they project directly to the external layer of the median eminence. If they do, the projection is sparse by comparison with the other systems mentioned above. The powerful effects of β-endorphin on reproduction are mediated probably by alterations in GnRH and prolactin secretion.

As may be seen from this brief survey of the parvocellular neurosecretory systems, a great richness of detail has accrued in a relatively short space of time due to the advent of powerful immunocytochemical techniques. However, it should be mentioned that this overall picture is only as accurate as the specificity of the antibodies used to reveal it, and it may change as they become more specific. These methods have also added a new and perplexing degree of complexity to the organization of this, and probably most neural systems, namely the *coexistence* of multiple synaptic and humoral messengers within the same neuron (Lundberg and Hökfelt 1983). This will briefly be summarized in a hypothalamic context below.

Patterns of Peptide, Amine and Amino Acid Coexistence Within Hypothalamic Neurosecretory Neurons

Coexistence of multiple messengers in neurons is demonstrated by examining thin, adjacent brain sections stained to reveal different antigens, or by sequential immunostaining of the two or more antigens, with an intervening elution step which removes the first primary antiserum prior to staining with the second (Tramu *et al.* 1978), or by the use of double staining methods in which two antigens can simultaneously be localized by visualizing their specific antisera differentially. Coexistence was earlier demonstrated in medullary 5-HT neurons which were shown to contain first Substance P and later TRH (the same TRH found in the hypothalamic paraventricular parvocellular neurosecretory neurons). Since then, the noradrenergic and adrenergic neurons of the medulla and pons, and the dopaminergic neurons of the substantia nigra and ventral tegmental area in the midbrain have all been shown to contain one or more neuropeptides (see Hökfelt *et al.* 1986 for review). These include NT and neuropeptide Y in the case of noradrenergic and adrenergic neurons (Everitt *et al.* 1984, Hökfelt *et al.* 1984a) and CCK and NT in the case of midbrain dopaminergic neurons (Hökfelt *et al.* 1980, 1986).

Patterns of some of the reported instances of the coexistence of peptides, amines and amino acids in hypothalamic neurons are shown in Table 1. Special mention will be made of some of them here. The oxytocin and vasopressin neurons of the PVN and SON contain a rich array of additional messengers which include opioid peptides, such as dynorphin and enkephalin, and also cholecystokinin and CRF. Parvocellular, CRF-containing neurons of the PVN have

Table 1. *Instances of the colocalization of peptide, amine and amino acid chemical messengers in hypothalamic neurons projecting to the external layer of the median eminence and neurohypophysis (abbreviations as in text)*

Messenger	Site	Coexisting messengers	
Dopamine	Arcuate nucleus	GRH	
		GABA	Enk
		NT	DYN
		GAL	
β-Endorphin	Arcuate nucleus	POMC	
		CLIP	
		ACTH	
		MSH	
GnRH	mPOA	LTC$_4$	GAP
CRF	PVN	AVP	
		OT	
		PHI (?)	
		ENK	
		NT	
		VIP	
TRH	PVN	ENK	
NT	PVN	ENK	
		CRF	
	Arcuate nucleus	DA	GRH
AVP	PVN/SON	DYN	
	Neurohypophysis	ENK	
		CRF	
		GAL	
OT	PVN/SON	DYN	
	Neurohypophysis	ENK	
		CRF	
		CCK	

been shown to contain AVP, particularly after adrenalectomy, when ACTH secretion is high (Tramu *et al.* 1983). These observations are relevant in the context of the long suspected role of AVP in the regulation of ACTH secretion.

Again in the PVN, the demonstration that nearly all CRF- containing neurons also contain PH 1 (a peptide closely related to vasoactive intestinal polypeptide and coded by the same gene) and also Met-enkephalin, suggests that the influence of these other peptides, alone or in combination with CRF, on ACTH secretion should be carefully investigated. Peptide HI has, in addition, been seen to release prolactin from pituitary cells in culture. Recently, VIP-immunoreactivity has been visualized in the same population of neurons, along with its mRNA, but only when rats are suckling their young.

In the medial preoptic area, GnRH neurons have

been shown to contain not only the precursor molecule of GnRH but also a 56 amino acid peptide possessing prolactin release-inhibiting activity as well as LH and FSH releasing activity (Philips *et al.* 1985). This intriguing combination of biologically active compounds within the same neurons has already attracted much functional interest. In addition, and particularly in the median eminence, GnRH has been seen to coexist with the leukotriene LTC$_4$. This compound has also been shown to affect GnRH secretion from the anterior pituitary.

The relatively small arcuate nucleus contains a bewildering number of instances of coexistence of peptides with other peptides, amine or amino acid (see Everitt *et al.* 1986 for review). For example, dopamine neurons have been shown additionally to contain either GRH, galanin, neurotensin, GABA and dynorphin. In addition, GRH, galanin and dopamine as well as GRF, neurotensin and dopamine have all been demonstrated within the same cells and all have very similar terminal domains in the external layer of the median eminence (Everitt *et al.* 1986).

At present little is known of the physiological actions, if any, of peptides such as galanin, PHI and neurotensin on the anterior pituitary. However, the fact that these peptides, as well as GABA and GRH, may be coreleased with dopamine into portal vessels suggests that the effects of combined manipulation of peptides and amines on adenohypophyseal secretion should be carefully studied (Everitt *et al.* 1986).

References

Barry J, Hoffman GE, Wray S (1985) LHRH-containing systems. In: Björklund A, Hökfelt T (eds) Handbook of chemical neuroanatomy, Vol 4, Part 1. Elsevier, Amsterdam

Card JP, Moore RY (1984) The suprachiasmatic nucleus of the golden hamster: immunohistochemical analysis of cell and fibre distribution. Neuroscience 13: 415–431

Conrad LCA, Pfaff DW (1976a) Efferents from the medial basal forebrain and hypothalamus in the rat. I. An autoradiographic study of the medial preoptic area. J Comp Neurol 169: 185–220

— — (1976b) Efferents from the medial basal forebrain and hypothalamus in the rat. II. An autoradiographic study of the anterior hypothalamus. J Comp Neurol 169: 221–261

Dahlstrom A, Fuxe K (1964) Evidence of the existence of monoamine-containing neurons in the central nervous system I. Demonstration of monoamines in the cell bodies of brain stem neurons. Acta Physiol Scand 62 [suppl] 232: 1–55

Everitt BJ, Hökfelt T, Meister B, Melander T, Terenius L, Weir JY, Goldstein M, Cuello C, Elde R, Vale W, Dockray G, Rokaeus A, Theodorsson-Norheim E, Edwardson J, Greengard P, Walaas I, Ouimet C, Hemmings H, Weber E (1986) Immunohistochemistry of transmitters, peptides and DARPP-32 in the hypothalamic arcuate nucleus-median eminence complex with special

reference to coexistence in dopamine neurons. Brain Res Rev 11: 97–155

— — Terenius L, Tatemoto K, Mutt V, Goldstein M (1984) Differential coexistence of neuropeptide Y (NPY)—like immunoreactivity with catecholamines in the central nervous system of the rat. Neuroscience 11: 443–462

Fibiger HC (1982) The organisation and some projections of cholinergic neurons of the mammalian forebrain. Brain Res Rev 4: 327–388

Hamilton WJ, Mossman HW (1982) Human embryology, 4th Ed. Heffer, Cambridge, pp 437–535

Haymaker W, Anderson E, Nauta WJH (1969) The hypothalamus. Ch C Thomas, Springfield, Illinois

Heimer L (1983) The human brain and spinal cord. Springer, New York

Hökfelt T, Everitt BJ, Theodorsson-Norheim E, Goldstein M (1984a) Occurrence of neurotensinlike immunoreactivity in subpopulation of hypothalamic, mesencephalic and medullary catecholamine neurons. J Comp Neurol 222: 543–559

— — Holets UR, Meister B, Melander T, Schalling M, Staines W, Lundberg JM (1986) Coexistence of peptides and other active molecules in neurons—diversity of chemical signalling potential. In: Iverson LL (ed) Fast and slow chemical signalling in the nervous system. Oxford University Press, Oxford

— Johannson O, Goldstein M (1984b) Central catecholamine neurons as revealed by immunohistochemistry with special reference to adrenalin neurons. In: Björklund A, Hökfelt T (eds) Handbook of chemical neuroanatomy, Vol 2, Part 1. Elsevier, Amsterdam, pp 157–276

— Ljungdahl A, Lundberg JM, Schultzberg M (1980) Peptidergic neurons. Nature 284: 515–521

— Martensson R, Bjorklund A, Kleinau S, Goldstein M (1984) Distributional maps of tyrosine hydroxylase-immunoreactive neurons in the rat brain. In: Björklund A, Hökfelt T (eds) Handbook of chemical neuroanatomy, Vol 2, Part 1. Elsevier, Amsterdam, pp 277–379

Johansson O, Hökfelt T, Elde R (1984) Immunohistochemical distribution of somatostatin-like immunoreactivity in the central nervous system of the adult rat. Neuroscience 13: 265–339

Krieger DT, Hughes JC (1980) Neuroendocrinology. Sinauer Associates Inc., Sunderland, Massachussetts

Lechan RM, Nestler JL, Jacobson S (1982) The tuberoinfundibular system of the rat as demonstrated by immunocytochemical localization of retrogradely transported wheat germ agglutinin (WGA) from the median eminence. Brain Res 245: 1–15

Lightmann SL, Everitt BJ (Eds) (1986) Neuroendocrinology. Blackwell Scientific Publications, Oxford

Lindvall O, Bjorklund A (1978) Organisation of catecholamine neurons in the rat central nervous system. In: Iversen LL, Iversen SD, Snyder SH (eds) Handbook of psychopharmacology, Vol 9. Plenum, New York, pp 139–231

Lundberg JM, Hökfelt T (1983) Coexistence of peptides and chemical transmitters. Trends Neurosci 6: 325–333

Mesulam M-M, Mufson EJ, Wainer BH, Levey AI (1983) Central cholinergic pathways in the rat: an overview based on an alternative nomenclature (Ch 1-Ch 6). Neuroscience 10: 1185–1201

Millhouse OE (1969) A Golgi study of the descending medial forebrain bundle. Brain Res 15: 341–363

Moore RY (1978) The central nervous system and the neuroendocrine regulation of reproduction. In: Yen SSC, Jaffe RB (eds) Reproductive endocrinology, pp 3–34

— Bloom FE (1978) Central catecholamine neuron systems: the dopamine systems. Ann Rev Neurosci 1 Pp.

— — (1979) Central catecholamine neuron systems: anatomy and physiology of the norepinephrine and epinephrine systems. Ann Rev Neurosci 2: 113–168

Morgane PJ, Panksepp J (eds) (1979) Anatomy of the hypothalamus. Marcel Dekker, New York

Ouimet CC, Miller PE, Hemmings H, Walaas SI, Greengard P (1984) DARPP-32, a dopamine and adenosine 3′: 5′-monophosphate-regulated phosphoprotein enriched in dopamine-innervated brain regions. 3: immunocytochemical localization. J Neurosci 4: 111–124

Palkovits M (1982) Neuropeptides in the median eminence: Their sources and destinations. Peptides 3: 299–303

Phillips HS, Nikolics K, Branton D, Seeburg PH (1985) Immunocytochemical localization in rat brain of a prolactin release-inhibiting sequence of gonadotropin-releasing hormone prohormone. Nature 316: 542–545

Reichlin S, Baldessarini RJ, Martin JB (1978) The hypothalamus. Research Publications: Ass Res Nervous and Mental Disease 56. Raven Press, New York

Rhodes CH, Morrell JI, Pfaff DW (1982) Cytoplasmic peptide content and nuclear estrogen binding of magnocellular neurons in the hypothalamus of Long-Evans and Brattleboro rats. Ann NY Acad Sci 767–775

Saper CBL, Swanson LW, Cowan WM (1976) The efferent connexions of the ventromedial hypothalamus of the rat. J Comp Neurol 169: 409–442

Sawchenko PE, Swanson LW (1982) The organisation of noradrenergic pathways from the brainstem to the paraventricular and supraoptic nuclei in the rat. Brain Res Rev 4: 275–325

— — Vale WW (1984) Corticotrophin-releasing factor: co-expression within distinct subsets of oxytocin-, vasopressin- and neurotensin- immunoreactive neurons in the hypothalamus of the male rat. J Neurosci 4: 1118–1129

Sidman RL, Rakic P (1982) Development of the human central nervous system. In: Haymaker W, Adams RD (eds) Histology and histopathology of the nervous system. Ch C Thomas, Springfield, Illinois, pp 3–145

Sofroniew M (1985) Vasopressin, oxytocin and their related neurophysins. In: Bjorklund A, Hokfelt T (eds) Handbook of chemical neuroanatomy, Vol 4, Part I. Elsevier, Amsterdam

Steinbusch H (1981) Distribution of serotonin immunoreactivity in the central nervous system of the rat—cell bodies and terminals. Neuroscience 6: 557–618

Stumpf WE, Grant LD (eds) (1974) Anatomical neuroendocrinology. S Karger, Basel

Swanson LW (1976) An autoradiographic study of the efferent connections of the preoptic region in the rat. J Comp Neurol 167: 227–256

— Cowan WM (1975) The efferent connections of the suprachismatic nucleus of the hypothalamus. J Comp Neurol 160: 1–12

— Kuypers HGJM (1980) The paraventricular nucleus of the hypothalamus: cytoarchitectonic subdivision and organization of projections to the pituitary, dorsal vagal complex and spinal cord as demonstrated by retrograde fluorescence double-labelling methods. J Comp Neurol 194: 555–570

Szentagothai J, Flerko B, Mess B, Halasz B (1968) Hypothalamic control of the anterior pituitary. An experimental, morphological study, 3rd Ed. Akademiai Kiado, Budapest.

Takeda N, Inagaki S, Shiosaka S, Taguchi Y, Oertel WH, Tohyama M, Watanabe T, Wada H (1984) Immunohistochemical evidence for the coexistence of histidine decarboxylase-like and glutamate decarboxylase-like immunoreactivities in nerve cells of the posterior hypothalamus of rats. Proc Natl Acad Sci USA 81: 7647–7650

Tramu G, Croix C, Pillez A (1983) Ability of the CRF immunoreactive neurons of the paraventricular nucleus to produce a vasopressin-like material. Neuroendocrinology 37: 467–469

— Pillez A, Leonardelli J (1978) An efficient method of antibody elution for successive or simultaneous localization of two antigens by immunochemistry. J Histochem Cytochem 26: 322–324

Van den Pol A (1982) The magnocellular and parvocellular nucleus of rat: intrinsic organization. J Comp Neurol 206: 317–345

Correspondence: Dr. Barry J. Everitt, Department of Anatomy, University of Cambridge, Downing Street, Cambridge CB2 3DY, England.

Acta Neurochirurgica, Suppl. 47, 16–37 (1990)

Neurohormonal Communication in the Brain

J. D. Vincent and **G. Simonnet**

INSERM U. 176, Domaine de Carreire, Bordeaux, France

And the Lord said, "Behold, they are one people and they have all one language" ... Therefore its name was called Babel because there the Lord confused the language of all the earth and from there the Lord scattered them abroad over the face of all the earth.

Genesis

Introduction

Why are there tens of chemical messengers, when just two—one stimulatory the other inhibitory—would suffice for communication between the many different types of neuron, that act as mediators of the unique signal—the action potential? A naive question with a naive answer: there are several messengers because there are several types of message to be delivered. These latter aren't limited simply to the opening of ionic channels gathered in a small area of the neuronal membrane to produce a localised depolarisation or hyperpolarization (excitatory or inhibitory post-synaptic potentials), but consist of complex modifications bearing on the whole cell thanks to the intervention of an intracellular second messenger.

The neuron theory based on Sherrington's concept (1906) states that the nerve cell forms a "communication unit" dynamically polarised between a reception zone (dendrites) and a transmission zone (axons). The signal transmitted by the axon—the action potential obeying an all or nothing law—is the result of algebraic summation of the inhibitory and excitatory post-synaptic potentials; it in turn stimulates the liberation at axon terminals of one type of mediator (Dale's principle, 1935). During the last two decades certain facts have contradicted the dogmatic character of this theory: different regions of the nerve cell show different degrees of excitability and the variation in local currents

creates a veritable spatio-electric geometry in the neuron quite different from the classical polarisation. The sodium-dependant action potential has lost its status as an "all or nothing" unique and universal signal. Calcium dependant action potentials with variable activation levels are found throughout certain dendritic regions and taken with the influxes responsible for message transmisson, provide the electrical basis for modulation of the incoming signal (Llinas *et al.* 1984).

Modulation is a term which came into wide use ten years ago, as a semantic "junk box" in order to mask the perplexity of neurobiologists faced with an ever increasing "babelisation" of the brain. Not only was there multiplication of the number of messenger candidates and their effects, but it also appeared that the same neuron could speak several languages or was at least capable of liberating several substances which could be called on to serve as neurotransmitters. The axon terminals no longer constituted the exclusive liberation site, nor the synaptic cleft the sole action point of the mediator (Emson 1985).

This review does not seek to add an nth contribution to semantic dicussions on the respective meanings of neurotransmisson and neuromodulation (Dismukes, 1979). It has become quite common to argue over words whose meanings we ignore, and to start a nomenclature before starting to understand. The definition of a hormone is, on the other hand, entirely unambiguous: it is any chemical messenger (from the greek *ormao*, I excite, I awaken) produced by a cell whose site of action is exterior to the cell. An indispensable addition to this definition requires that the substance should act at a distance from its production site. The notion of a hormone thus implies the *diffusion* of the message in a communication *space* which may be to the blood, other body fluids or just the extracellular space. *Stricto sensu* the NEUROHORMONES are chemical messengers, secreted by neurons, whose effects will take place distant to their site of liberation.

The chemical nature of the substance does not pose a problem—it can be an amine just as well as a peptide. The liberation site is not exclusive either—it may be at a differentiated membrane or at a non specialised region, as long as its effect isn't limited to the postsynaptic territory opposite. The *diffusion of the message* to its receptors can take place via the blood, the cerebrospinal fluid or the extracellular space, but could equally as well be carried out in the intracellular space of the target cell via a second messenger. Under these conditions, the central nervous system can be considered as an endocrine gland, either liberating its hormones into the circulatory system by neurohaemal organs, or reserving them for its own internal use. We are conscious of the insufficiency of such an interpretation of the term "neurohormone". Its one merit is that it allows us to review the vast catalogue of substances secreted by neurones and their effects; our ambition lies not in a classification and an exhaustive study but simply to a description of certain elements in no particular order whilst awaiting novel concepts for a study of interneuronal communication.

The Neuroendocrine Systems

During the last decade, the neurons have lost the exclusive right to electrophysiological properties; numerous endocrine cells have been shown to have comparable membrane characteristics. These observations have led to a revision of the classical opposing positions of the nervous and endocrine systems (Fig. 1). New schemes of organisation have been proposed which attempt to integrate the similarities between the endocrine and nervous cells. These classifications take into consideraton the finding that the same substances are found in the two cell types. This histological ubiquity is mirrored by an evolutionary ubiquity, both ontogenetic and phylogenetic.

A) Nervous Cells and Endocrine Cells

1) The APUD Cells

Pearse (1966a), observing the property or at least the capacity of certain peptide secreting endocrine cells to produce the biogenic amines, proposed to group them under the generic appellation of *APUD system*: Amine Precursor Uptake Decarboxylase.

Pearse later modified his initial concept when he described the *Diffuse Neuroendocrine System* (DNES)

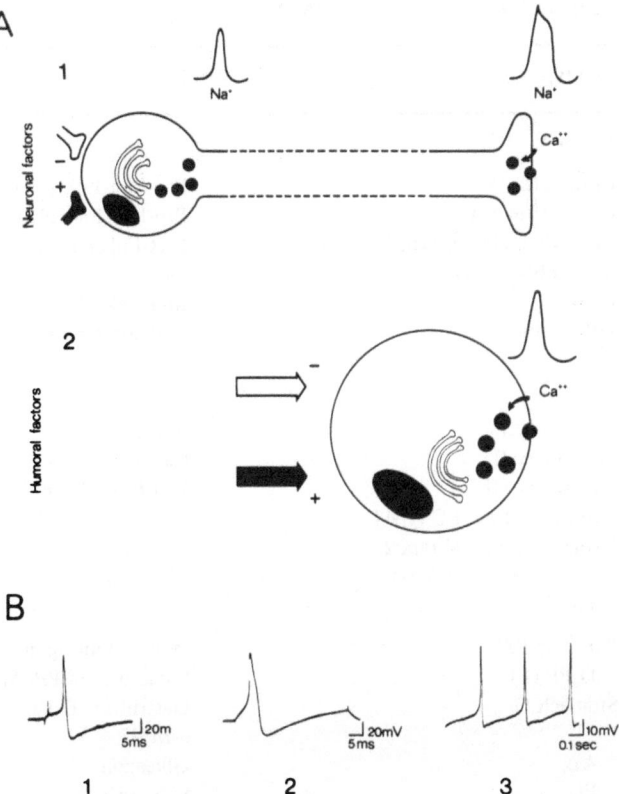

Fig. 1. A) Schematic representation of a neuron (*1*) and of an endocrine cell (*2*) during stimulus-secretion coupling. B) Examples of action potentials: *1* mouse spinal neuron in primary culture; *2* mouse dorsal root ganglion neuron in primary culture; *3* rat anterior pituitary prolactin cell

which groups together cells whose embryological origins may differ but which share a common neuroendocrine program.

The DNES can be divided, according to Pearse (1983), into a central and a peripheral division. Most cells of the latter group are localised in the gastrointestinal tract with inclusion of the pancreas (see Table 1). For the first group, an origin in the neural crest is generally accepted (Le Douarin 1982); the origin is disputed, or less well established for cells of the second group (Pearse 1983). It has been shown, on the other hand, that cells from the third group are not derived from the neural crest (Le Douarin and Teillet 1973). Their origin remains mysterious for the time being.

The *central division* according to Pearse (1983) (Table 2) regroups the cells of the adenohypophysis, the pineal gland and the parvo and magnocellular systems in the hypothalamus. Each of them have the capacity or at least the potential, to synthesise peptides and amines (Pearse 1969). Depite the current view of

Table 1. *Peripheral division of DNES*

Cell type	Peptides	Amines
Group 1		
Chromaffin A	Pro-Met-Enkephalin	A
Chromaffin NA	Pro-Leu-Enkephalin	NA
Carotidian glomus type 1	Met-Enkephalin	DA
Melanoblastocyte	—	(Cysteinil-DOPA)
Merkel	Met-Enkephalin	—
UB thyroïd	Calcitonin Somatostain	5-HT
Group 2		
Pulmonary I	Bombesin	—
Pulmonary II	Calcitonin	5-HT
Pulmonary III	Leu-Enkephalin	—
Urogenital tract EC type 1	—	5-HT
Urogenital tract U type 2	—	—
Group 3		
Pancreas BA	Insulin, Glucagon	5-HT
D PP (F)	Somatostatin PP, Met-Enkephalin	DA
Stomach G	Gastrin-17, Enkephalin ACTH	
	α-MSH	—
AL	Glucagon	—
EC	Substance P	5-HT
ECL	—	H
D	Somatostatin	—
GI tract EC	Substance P	5-HT, MT
M	Motilin	—
L, S	Glycentin, Secretin	—
I, P	CCK, Bombesin	—
D, K	Somatostatin, GIP	—
N	Neurotensin	—
H	VIP, PHI	—
IG	Gastrin 34	—
TG	CCK	—

a stomoectodermal origin for the adenohypophysis, certain authors have advanced the hypothesis of a neuroectodermal origin for the whole hypothalamo-hypophyseal complex (Takor-Takor and Pearse 1975, Nakajima *et al.* 1980, Cocchia and Miani 1980).

Bearing in mind Pearse's definition—the capacity of the APUD cells to synthesise peptides and amines at the same time-it seems that we can expand this framework of the central and peripheral divisions of the DNES to encompass the whole of the nervous system. Recent literature shows numerous examples of the coexistance of biogenic amines and peptides in cells of the autonomic nervous system and in the central nervous system (Table 3). The central catecholaminergic neurones might be considered as sympathetic ganglia translocated to the brain (Emson 1985).

This definition of the APUD and DNES systems might seem restrictive as it does not take into account the acetylcholine-peptide coexistence observed in numerous parasympathetic and central cells. During perinatal development, *in vivo* (Hill and Hendry, 1977) the sympathetic cells remain potentially capable of synthesising acetylcholine and noradrenaline at the same time, before the intervention of perinatal differentiation factors which orientate the system one way or the other.

The embryological non differentiation of noradrenergic and cholinergic neurones thus allows us in theory to class them within the APUD cell framework. At this point it may be preferable to enlarge the conceptual outline defined by Pearse to engulf all cells capable of producing at the same time a substance identical or similar to a classical neurotransmitter and a peptide

Table 2. *Central division of DNES*

Cellular type	Peptides	Amines
Pineal	AVT LRH	MT 5-HT
Hypothalamic m.c.	AVP AVT	DA NA 5-HT
Hypothalamic p.c.	RFs RIFs	DA NA 5-HT
Adenohypophysis	FSH LH TSH	DA
	GH PRLAVTH	NA
	MSH β-LPH	5-HT
	β-endorphin	H
	Gastrin Calcitonin	(T)
Hypophysis: intermediary lobe	ACTH MSH	
	β-LPH	
	β-endorphin	DA NA 5-HT
	Calcitonin	

Symbols

m.c.: magnocellular	ECL: enterochromaffin-like
p.c.: parvocellular	PP: pancreatic polypeptide
AVT: arginin vasotocin	GIP: glucose dependant, insulin releasing
AVP: arginin vasopressin	peptide
LRH: luteinising releasing	VIP: vasoactive intestinal peptide
hormone	PHI: porcine intestinal heptacosa peptide
RFs: releasing factors	NA: noradrenalin
RIFs: release inhibiting factors	H: histamine
UB: ultimobronchial	T: tyramine
UG: urogenital	MT: melatonin
A: adrenalin	
EC: enterochromaffin	

Table 3. *Some examples of coexistence in the same cell of a biogenic amine with a peptide in the autonomic nervous system (from Osborne 1983) and the central nervous system (from Hökfelt et al. 1984) (for symbols see Table 2)*

Cellular type (animal species)	Peptides	Amines
Superior cervical ganglion (Rat)	Enkephalin	NA
Inferior mesenteric ganglion (Cat)	Somatostatin	NA
Perivascular nerves (Cat)	Avian pancreatic polypeptide	NA
Ventral tegmental area (Rat)	Neurotensin	DA
Ventral tegmental area (Rat, Man)	Cholecystokinin	DA
Locus coeruleus (Rat)	Enkephalin	NA
Locus coeruleus (Rat)	Neuropeptide Y	NA
Medulla oblongonta (Rat)	Neuropeptide Y	A
Medulla oblongonta (Rat)	Neurotensin	A
Medulla oblongonta (Rat)	Substance P	5-HT
Medulla oblongonta (Rat)	TRH	5-HT
Medulla oblongonta (Rat)	Enkephalin	5-HT

substance showing hormonal properties. The para-neurone concept defined by Fujita (1977) is compatible with this definition.

2) Neurones and Paraneurones

Fujita notes that "the neurones which liberate their secretory products—peptides, amines, acetylcholine and nucleotides—into intracellular spaces or into blood capillaries are identical in nature to a group of endocrine cells called *paraneurones* which produce the same type of products" (Fujita *et al.* 1984). Whilst the APUD cell concept is based on biochemical criteria, that of the neurone/paraneurone distinction is principally dependant on cytomorphological data.

a) Secretory granules and synaptic vesicles.

In comparison with secretory granules, found in the paraneurones, the granules observed in the magnocellular neurones (180–200 nm) and the large dark centred synaptic vesicles (90–120 nm) present in certain neurones, are not fundamentally different. The granules and vesicles contain Ca^{2+} and ATP in addition to a number of peptides.

The small light or dark—cored synaptic vesicles (30–50 nm) don't generally seem to contain peptides. The small dark cored vesicles are characteristic of adrenergic terminals. The light vesicles belong to cholinergic terminals. Whilst amines and peptides can coexist in the same vesicle (Pelletier et al. 1981) the presence of acetylcholine and a peptide in a terminal necessitates separate vesicular storage (Lundberg 1981).

b) Granule dynamics in neurones and paraneurones.

The stages of formation, maturation and liberation of the granules in chromaffin cells, adenohypophyseal cells, and pancreatic β cells and other paraneurones have been the object of diverse morphological studies (Kobayashi 1977, Fujita 1984). The proteins, enzymes and hormone precursors, after synthesis in the endoplasmic reticulum, are transfered to the Golgi apparatus to be packaged in the form of secretory granules. The incorporation of amines and ATP is carried out later during maturation (Fig. 2). The same scheme holds for neurones in which the maturation and amine incorporation take place during transport of the neurosecretory granules or the large dark cored vesicles (Morris et al. 1978).

Finally, it would seem that a unitary theory is possible concerning granules and vesicles. Apparently a simple quantitative difference permits the distinction between secretory granules and synaptic vesicles. As to the contents, the peptides are packaged at the moment of bag formation and the other messenger substances, amines, acetylcholine and ATP are incorporated later on.

3) Electrical Excitability of Neurones and Para-neurones

An important consequence of their embryological and functional relationship is that paraneurone type endocrine cells and neurones share similar electrophysiological properties. Some endocrine cells which respond to stimuli by action potentials include chromaffin cells (Biales et al. 1976), adenohypophyseal cells (Vincent et al. 1985), pancreatic cells (Matthews and

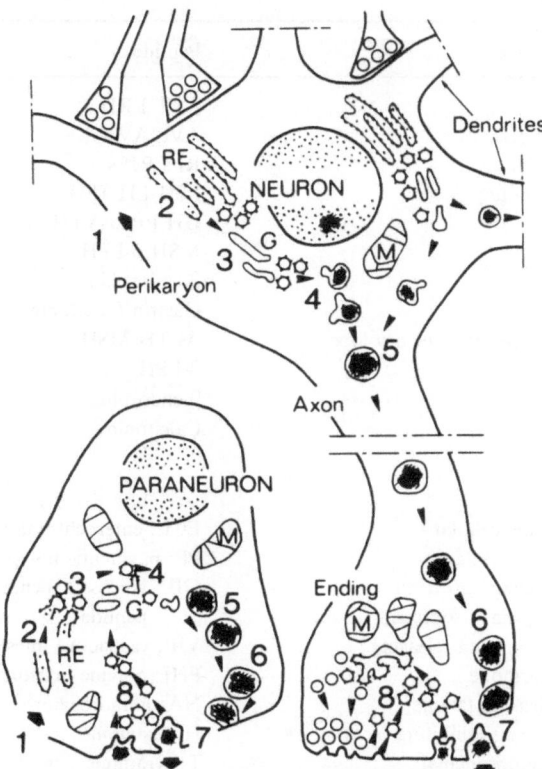

Fig. 2. Schematic representation of granule dynamics in a neuron and a paraneuron. *1* Uptake of extracellular material; *2* ribosomal protein synthesis followed by segregation within the endoplasmic reticulum; *3* transport of protein to the Golgi apparatus; *4* formation of secretory granules; *5* maturation of the newly formed secretory granules with incorporation of substances such as monoamines, ATP and CA^{2+}; *6* stockage of secretory granules; *7* exocytosis; *8* formation of coated vesicles and synaptic vesicles. *RE* endoplasmic reticulum; *M* mitochondria; *G* Golgi apparatus. (From Fujita et al. 1984)

Sakamoto 1975a and b). We would emphasize, without going into the details of the mechanisms, the role of voltage sensitive Ca^{2+} channels in stimulus secretion coupling in these various paraneurones (Vincent et al. 1985).

B) Histological Ubiquity of the Neurohormones

Neurohormones are omnipresent in the organism. The demonstration of this finding is due in large part to the extraordinary methodological progress made in the last ten years. Immunohistochemical methods have revealed the diversity and extent of the localization of nervous and endocrinological peptides from an ever expanding catalogue (Hökfelt et al. 1984). The existence of a specific reuptake system in the nerve terminals facilitates tracing monoaminergic pathways by autoradiography (Descarries and Beaudet 1983). The concept of the neuropeptide has been developed in the light of the growing discovery of the presence of peptide hormones inside the brain. First it was the hypothalamic hormones, TRH and LH-RH, isolated by Guillemin and Schally, that were found in the central nervous system outside the hypothalamus (Berson and Yalow 1973). The hormones of neural origin, vasopressin and oxytocin, were shown to be in the fibres and axon terminals situated in the brain and the spinal cord (Buijs *et al.* 1978). The brain also contains adenohypophyseal hormones as first shown for corticotrophin hormone (Liotta *et al.* 1980). The discovery by Vanderhaegen *et al.* (1975), of the presence of gastro-intestinal hormones—or at least of substances displaying an immunoreactivitiy to antisera directed against these hormones—was the first demonstration, rapidly followed by others, of the presence of a gastro-intestinal hormone in the brain. Inversely, peptides first discovered in the brain, such as the enkephalins (Hughes *et al.* 1975) and others derived from the pro-enkephalins A and B have been largely rediscovered in the hypophysis, the adrenal medulla, the sympathetic ganglia and the gastro-intestinal tract.

Table 4 shows an incomplete list of the principal neuropeptides present in the central nervous system. Some of the peptides (such as gastrin, CCK, VIP) are present both in endocrine cells *stricto sensu* and in nerve cells. Other peptides are exclusively found in digestive endocrine cells, such as secretin, motilin, insulin and glucagon. Finally there are some that have only been observed in the nerve cells of the gut lining, such as bombesin, the enkephalins and substance P. We find hormonal peptides (6) in the miscellaneous group whose actions in the central nervous system are indisputable but which remain difficult to classify. This is the case for instance of angiotensin II, a circulatory hormone whose effects are spectacular after intracerebral injection but whose presence in the brain remains controversial (Simonnet *et al.* 1984); this is also true for atriopeptin (an atrial natriuretic factor) present both in the circulatory system and in the brain (Standaert *et al.* 1985).

One cannot but be astonished at the low concentrations of the peptides in the brain (10^{-12} to 10^{-15} moles per milligram of protein), compared with the high concentrations of these hormones in other tissues such as the digestive tract. This difference does not minimise their role in the central nervous system and may be justified by a smaller diffusion volume in the brain than in the rest of the body where they are subject to dilution in the circulatory system before reaching their target cells (Krieger 1983).

Table 4. *Peptidergic neurohormones*

-1- Hypothalamic hormones

Thyroïd Releasing Hormone (TRH)
Luteinising Hormone Releasing Hormone (LH-RH)
Somatostatin (SRIF)
Corticotrophin Releasing Hormone (CRH)
Growth Hormone Releasing Hormone (GH-RH)

-2- Neurohypophysial Hormones

Vasopressin (AVP or ADH)
Oxytocin
Met-Enkephalin
Dynorphin 1–8

-3- Adenohypophyseal Peptides

Adreno Cortico Trophic Hormone (ACTH)
β-endorphin
α-Melanocyte Stimulating Hormone (αMSH)
Prolactin
Luteinising Hormone (LH)
Growth Hormone (GH or STH)
Thyroid Stimulating Hormone (TSH)

-4- Peptides isolated from invertebrates

Hydra head activator
Mollusc cardiac-excitatory peptide

-5- Gastroinestinal peptides

Vasoactive Intestinal Polypeptide (VIP)
Cholecystokinin (CCK)
Gastrin
Substance P
Neurotensin
Met-enkephalin
Leu-enkephalin
Insulin
Bombesin
Secretin
Somatostatin
TRH
Motilin
Avian Pancreatic Polypeptide (APP)

-6- Miscellaneous

Angiotensin II (AII)
Atriopeptin (NAF)
Bradykinin
Carnosin
Sleep peptide(s)
Calcitonin
Calcitonin-gene related peptide (CGRP)
Neuropeptide Y (NPY)
Polypeptide YY (PYY)

C) Evolutionary Ubiquity

Substances responsible for communication are present in living things even before organ differentiation. Hormones and neurotransmitters precede the appearance of endocrine and nervous systems. One could almost say that the mediators precede the appearance of life. The steroid hormones—œstrogens and corticoids—are synthesized in yeasts which also possess the corresponding receptors. A factor has been isolated from yeast membranes which, after recognition by a specific receptor, leads to reciprocal attraction between two partners. It so happens that this peptide (factor α) has an amino-acid sequence very similar to that of luliberin (LH-RH) whose receptor it even recognises; it is also capable of stimulating the reproductive function in mammals (Loumaye et al. 1982). Even at this early evolutionary stage there are thus substances: steroids, peptides—which will later become hormones—already dictating cell development, reproduction and relationships with a protective support.

Hormones or neuromediators—at this stage the difference has no meaning—are present in protozoans. In Tetrahymena we find endorphins, insulin, somatostatin, vasopressin and ACTH, or at least immunologically similar substances (Le Roith et al. 1982, Schwabe et al. 1983). These substances, secreted by the unicellular being, can act at a distance on other individuals—pheromone type action—or on the cell itself by binding to autoreceptors—autocrine type action. All such systems will be found operating in evolved animals and in man (Fig. 3). The central nervous system and the digestive tube, two routes in continuity with the outside world, are places of predilection for secretion of substances reponsable for communication. The nervous system is firstly neuroendocrine in the wide sense of the term, the secreted substances acting in the immediate vicinity (synapse) or at a distance from their place of fabrication (hormone). When a new substance appears on the evolutionary tree, it is seen first in the nervous system. The ancestral gene is expressed in neurons before its descendants manifest themselves in digestive cells. The formation of nerve centres—gathering of nerves and synapses—in worms precedes that of endocrine glands. The existence of endocrine glands quite independant from the nervous system is a relatively recent evolutionary event, characteristic of vertebrates and superior invertebrates (Krieger 1983).

It is possible for the same peptide to come into play in different ways in the course of evolution (Fig. 3). These roles may coexist within a given species. This is true for instance, of derivatives of pro-opiomelanocortin: a neuronal messenger in the brain, neuroendocrine in the hypothalamo-hypophyseal system, endocrine in the hypophysis and paracrine in the digestive and reproductive tracts (Larsson 1980, Tsong et al. 1982). Finally, a particular cell, such as the pinealocyte, can during evolution lose its nervous function (photoreception) to adopt the attributes of a purely endocrine secretory gland.

The same ubiquity, although less well researched, is found throughout ontogeny. The observations of Teitelman (1981) showed that mouse embryological pancreatic cells possess transitory tyrosine-hydroxylase activity from the 11th to the 15th day of embryological life that precedes the appearance of glucagon followed by insulin synthesis on the 12th and 15th days respectively. Under these conditions, the expression of a catecholaminergic phenotype would precede that of the peptidergic phenotype. These results can be linked with the observation, now well established, that certain ectodermal secretory cells of a metazoan (Laomedea flexuosa) contain dopamine (Knight 1970). In ontogeny, as in phylogeny, amines would thus precede peptides.

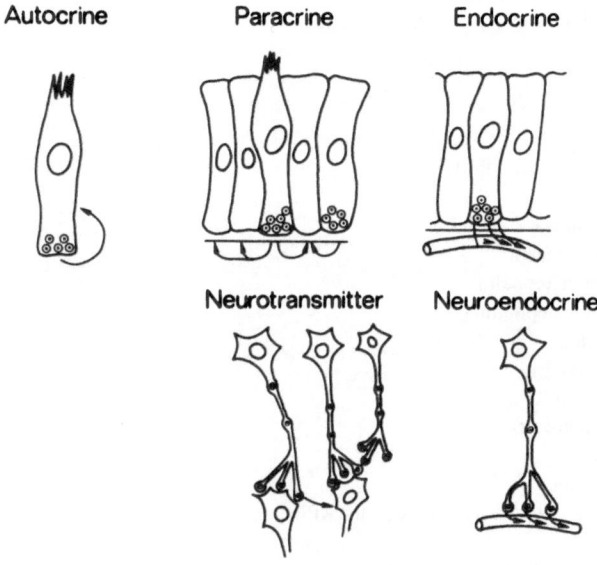

Autocrine   **Paracrine**   **Endocrine**

Neurotransmitter   **Neuroendocrine**

Fig. 3. Different modes of peptidergic secretion. In autocrine secretion, the product acts locally on its productory cell. In paracrine secretion the product acts on neighbouring cells by local diffusion within an extracellular medium or by "gap" junctions. In endocrine secretion, the product is transported to target cells by the blood. A neurotransmitter is liberated from axon terminals, it reaches the post synaptic site opposite or diffuses to neighbouring neurons; it can also be liberated at the contact of an axon terminal. In neuroendocrine secretion stricto sensu, the secretion product liberated by an axon terminal reaches is target via the circulatory system. (From Krieger 1983)

D) Functional Ubiquity

The major homeostatic systems all rely on 2 modes of intervention, one behavioural, the other metabolic. For example, to combat the cold and avoid a decrease in body temperature, an animal finds shelter, bristles its fur or feathers (behavioural mechanisms) and increases cell heat production (metabolic means). To compensate for water loss and to maintain the degree of body hydration, the individual drinks (behaviour) and decreases water elimination by the kidney (metabolism). To keep his weight constant, he eats (behaviour) and mobilises his reserves (metabolism). In the domain of species conservation, reproduction also demonstrates a harmonious combination of cellular phenomena (maturation of sexual cells) and behavioural means. Finally, in the rearing of the young, maternal behaviour is associated with milk production. Division between the brain—behaviour—on one side, and hormones and glands—metabolism—on the other, is inconceivable. Scrutiny of the chemical mechanisms reveals that the same substances intervene in the behavioural response as in the metabolic response. The substance acts just as well in the blood as a hormone as in the brain as a neurohormone.

For reproduction, LH-RH or luliberin, a hypothalamic hormone which by the intermediary of the hypophysis regulates sex cell maturation and release, when injected into the brain stimulates the reproductory behaviour sequence. Luliberin is also present in the ovary where it has a role as a local hormone.

Angiotensin II, which in the circulatory system provokes blood vessel contraction, is also present in the brain where it not only stimulates drinking behaviour, but also intervenes in nervous regulation of arterial pressure and is responsible for liberaton of antidiuretic hormone (Simonnet et al. 1979). In its adaptation of the contents to the container (drinking) or of the container to the contents (vasoconstriction), the ubiquity of the substance is astonishing, as hormone in the blood, neurotransmitter in ganglia of the sympathetic nervous system, and again as neurotransmitter or neurohormone at different levels of the central nervous system. The same dual destiny exists for the digestive hormones, insulin and gastrin, which are also found in the brain participating in the behavioural mechanisms of feeding.

What we perceive is that there is cerebral organisation of the coordinated mechanisms responsible for the regulation of the major functions necessary for survival of the individual and of the species. In the same way as the formation of an milieu interieur, in multicellular organisms necessitated the creation of support organs with precise functions, the appearance of a cerebral medium requires the formation within the brain of complex regulatory units whose master control is assured by the same chemical entities as those used at the peripheral organ level. In the same way as the homeostatic functions of the internal medium are served by coodination between specialised organs, liver or kidney, so the homeostatic units in the brain start with precise localisation in the centres—hunger centre, reproductive centre or respiratory centre as they are called—to extend to dispersed territories which sometimes cover the whole of the encephalus. On both sides of the blood-brain barrier, the same liaison agents thus have a double regulatory role. We shall try to demonstrate this functional ubiquity by the example of certain neurohormones.

1) Angiotensin II

The octapeptide angiotensin II (AII) is a powerful vasoconstrictory hormone liberated in response to disturbances of salt and water metabolism. In parallel with this well studied systemic intervention, it has been shown that AII can act directly on the brain inducing, according to the injection site, drinking behaviour (Fitzsimons 1972), hypertension (Ganten et al. 1978) and vasopressin release (Simonnet et al. 1979). Two pharmacologically distinct receptor types for AII can be identified on central nervous system neurones either by binding studies (Simonnet and Vincent, 1982, Laribi et al. 1985) or by electrophysiology (Legendre et al. 1984). Iontophoretic application of this peptide indicates the existence of selective sensitivity in certain regions. These are the septum (Simonnet et al. 1980, Huwyler and Felix 1980), the lateral hypothalamus (Wayner et al. 1973), the subfornical organ (Felix and Akert 1974), the hypothalamic magnocellular nuclei (Akaishi et al. 1980), the preoptic area (Gronan and York, 1978, Simonnet et al. 1980) and the midbrain (Suga and Suzuki 1979).

The presence of this peptide in the brain, on the other hand, has been less well demonstrated. Immunohistochemical methods have revealed an "angiotensin-like" reactivity in different regions of the central nervous system in particular in the lateral sympathetic columns of the rat spinal cord (Fuxe et al. 1976). By chromatography, Simonnet et al. (1984) showed that authentic endogenous AII is undetectable in the brain where a molecule of higher molecular weight exists, recognizable by immunological dosage. Analysis of the transformation of AII precursor in the presence of brain tissue supports the hypothesis that the rapidity of AII degradation in the brain could be the reason why it is not detectable.

AII's presence is suspected in preganglionic sympathetic neurones (Fuxe *et al.* 1976) and intervenes at the ganglionic relay level (Dun *et al.* 1978). Its action on vascular smooth muscle fibres completes its powerful vasoconstrictory actions. Finally our attention is drawn to the multiple roles played by this same substance, supported by different mechanisms at different levels of the organism—kidney, vessels, ganglion, spinal cord and brain—to bring about cooperation in a unique function: liquid homeostasis in the organism.

2) Atriopeptin

The heart is not only a simple hydraulic pump, but also an endocrine gland whose atrial myocytes secrete a peptide hormone: *atrial natriuretic peptide (ANP) or atriopeptin.* This peptide of 28 amino acids formed from a prohormonal precursor of 152 amino acids has been isolated in purified form from myocardial extracts or from plasma (Schwartz *et al.* 1985). It is liberated by the influence of a distension mechanism in the left auricle activated by an increase in blood volume.

ANP has specific receptors in the kidneys, the adrenal glands and the blood vessels. Its injection into an animal is followed by a diuresis and natriuresis leading to a spectacular reduction of blood volume. The biological effects of ANP are thus diametrically opposed to those of AII. As is the case for AII, ANP is found within the central nervous system. Atriopeptin has been recognised by immunohistochemistry and by immunological methods in hypothalamic extracts

(Tanaka *et al.* 1984). All the same, as for AII, the identification in nervous tissue has not been achieved. However, the presence in hypothalamic extracts of atriopeptin mRNA as revealed by Northern blot confirms the existence of endogenous atriopeptin in the brain.

It has been shown by immunohistochemistry that cerebral atriopeptin is localised in structures concerned with cardio-circulatory regulation (Fig. 4) (Saper *et al.* 1985). The largest group of immunoreactive neurones is situated in the anterior lining of the 3rd ventricle, an area called "AV 3 V". The destruction of this zone brings about chronic hypodypsia and hypernatremia (see the review of Standaert 1985). The organum vasculosum laminae terminalis in this region is devoid of a blood-brain barrier. Its neurones are connected to the paraventricular nucleus. The injection of atriopeptin into the 3rd ventricle blocks the spontaneous liberation of vasopressin and inhibits the behavioural (drinking) and hormonal (vasopressin release) responses to osmotic stimulation or injection of AII (Samson 1985, Antunes-Rodrigues 1986). Nerve terminals immunologically reactive to atriopeptin are also found in the region of the nucleus tractus solitarius, an area of the brainstem which receives information from the cardiovascular system, and in the nucleus parabrachialis which constitutes their first relay on the way to overlying cerebral structures (for review see Standaert 1985).

The same conclusion is reached for atriopeptin as for AII. This substance intervenes at different levels in the systematic organisation and in the central nervous system by a multitude of means to achieve a single homeostatic end which aims to avoid an excess of the circulating blood volume. AII and atriopeptin thus represent opposite poles of a single regulatory function: hydromineral equilibrium in the organism.

3) Vasopressin

In contrast to the two preceding cases (AII and ANP), vasopressin was first discovered as a hormone secreted by neurones of the hypothalamic magnocellular system (supraoptic and paraventricular nuclei), whose systemic actions are directed at the kidney and the blood vessels. Immunohistochemical studies have revealed more recently the presence of AVP in neurons situated outside the magnocellular system, and that nerve terminals containing vasopressin are widely distributed in the brain (septum, amygdala, habenula) in the brainstem (nucleus tractus solitarius and dorsal nucleus of the vagus) and in the spinal cord (for review see Sofroniew 1985).

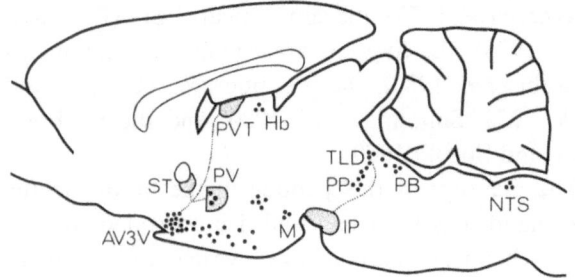

Fig. 4. Principal groups of neurons and nervous fibres showing immunoreactivity to atriopeptin in the rat. Abbreviations: *AV 3 V* anteroventral region of the 3rd ventricle; *ST* stria terminalis; *PV* paraventricular nucleus of the hypothalamus; *Hb* habenula; *TLD* laterodorsal tegmental nucleus; *PB* parabrachial nucleus; *PP* pedunculopontic nucleus; *IP* interpeduncular nucleus; *M* mamillary bodies; *NTS* nucleus tractus solitaris

The possible role of cerebral vasopressin in memory processes has been much debated. From their observations on the rat, De Wied (1965) and Van Wimersma Greidanus (1982) have postulated that vasopressin is involved with learning. This hormone may facilitate the consolidation, storage and learned response recall at the cerebral level. This hypothesis has proved controversial based in particular on the observation that vasopressin acts mainly on avoidance behaviour by means of its systemic effects, notably on arterial pressure (Le Moal 1981). Without entering the debate, it seems more important to us to show that cerebral vasopressin intervenes as the peripheral hormone in the case of body fluid homeostasis. Demotes-Mainard *et al.* (1986) have studied the liberation of intracerebral AVP in the anaesthetized rat using "push pull" microcannulae implanted in different regions of the brain (Rodriguez *et al.* 1983). They obeserved a basal liberation of vasopressin in the lateral septum and in the 3rd ventricle which is greatly increased by an intraperitoneal injection of hypertonic saline (Fig. 5). This response to a systemic osmotic stimulation is blocked when the perifusion liquid in the microcannula is calcium-free (in the presence of 0.1 mM EGTA). No basal release or release in response to an osmotic stimulation is observed in nervous structures such as the caudate nucleus (Fig. 5). One and the same systemic osmotic stimulation is thus capable both of stimulating vasopressin release into the blood via the posterior hypophysis and in the brain via the axon terminals projecting into the septum. The liberation of vasopressin into the cerebrospinal fluid of the 3rd ventricle in response to an osmotic stimulus has been observed by many other experimentors (Barnard and Morris 1982, Doris and Bell 1984, Epstein *et al.* 1983, Robinson and Jones 1982, Szczepanska-Sadowska *et al.* 1983, Wang *et al.* 1982, Zerbe and Palkovits 1984). We shall came back later to the role of cerebrospinal fluid as a neurohormone diffusion route in the interior of the central nervous system.

A hypovolemic stimulus is also capable, although to a much lesser extent, of provoking vasopressin liberation in the septum and in the cerebrospinal fluid (Demotes-Mainard *et al.* 1986). It thus appears to be indisputable that homeostatic processes implicated in hydromineral equilibrium are concerned simultaneously with vasopressin liberation in the blood and in the central nervous system. The septum in particular is the cerebral structure implicated in the control of neurohypophyseal functions (Lubar *et al.* 1968, Sibole *et al.* 1971). The electrical stimulation of the septum modifies the electrical activity of neurons in the magnocellular sys-

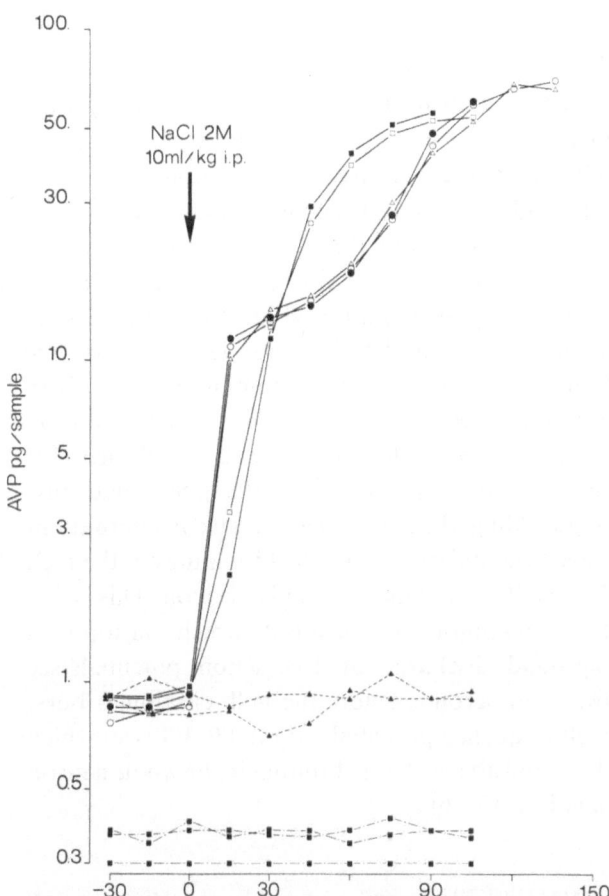

Fig. 5. Liberation of arginine vasopressin (*AVP*) in rat brain collected by the "push-pull" cannula technique. The basal liberation of AVP in the lateral septum (2–5 pg/fraction) can be stimulated by a peripheral osmotic load NaCl 2M, 10 ml/kg i.p.). This stimulation has no effect on AVP liberation in another central structure such as the caudate nucleus. Each curve represents one experiment as a function of time after the intraperitoneal injection of hypertonic saline solution

tem (Poulain *et al.* 1980, Shibuki 1984). The electrolytic destruction of the septum influences vasopressin liberation and drinking behaviour in response to an osmotic stimulation (Iovino *et al.* 1983). Conversely, electrical stimulation of the supraoptic nucleus, which increases septal vasopressin liberation (Demotes-Mainard *et al.* 1986) modifies the electrical activity of the septal neurones (Poulain *et al.* 1981). It must also be noted that iontophoretic application of vasopressin onto septal neurones provokes, depending on the authors, an excitation (Joels and Urban 1982) or an inhibition (Pestre *et al.* 1984). It therefore appears that the vasopressin liberated in the septum could play a feedback in controlling its own release and thus participate just as much as the peripheral hormone in hydromineral homeostasis.

4) Oxytocin

Oxytocin is known above all for its peripheral hormonal action. It is activated during lactation in the suckling response to initiate milk ejection.

The milk ejection reflex constitutes the best known model of a neuroendocrine reflex. The nerve centres responsible for hormone liberation (paraventricular and supraoptic nuclei) are well known and accessible to many experimental techniques. The oxytocinergic cells have been identified by immunohistochemistry and by their electrical activity. Their function during the milk ejection reflex is now well established (for review see Poulain and Wakerley 1982). Schematically, during suckling, the oxytocin is liberated in intermittent massive brief pulses (every 5 to 15 minutes in the rat), with each liberation inducing milk ejection. This is due to the activation of oxytocin cells which, on top of a background discharge of 1–3 action potentials/sec show, a few seconds before the milk ejection, a burst of high frequency potentials (up to 80–100/sec) which are brief and above all synchronous in the whole neuron population (Fig. 6).

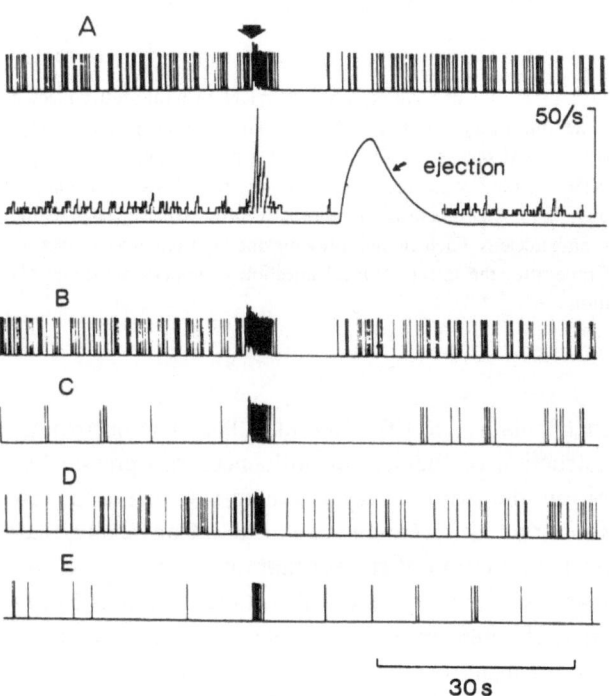

Fig. 6. A) Electrical activity recorded by an extracellular electrode from an oxytocin neuron of the supraoptic nucleus of a lactating rat. Note the burst of action potentials preceding milk ejection which is recorded by a pressure sensor placed in a nipple. In (B, C, D, and E) other examples of synchronous bursts recorded during a milk ejection reflex in smiliar experiments

The synchronous nature of the high frequency action potential burst throughout the oxytocinergic neurone population raises the question of basic mechanisms of this synchronisation. In our first study, we noticed that during lactation, the supraoptic nuclei were subject to morphological modifications. Normally the neurones are separated from one another by the glial elements. These elements shrink during lactation in such a way that the neuronal surfaces become adjacent. At the same time one may observe an increase in attachment plates between neurons. What is more there is increase in the number of presynaptic elements making simultaneous contact in the same section plane with two post synaptic elements. These two morphological factors may have important functional consequences, the neuronal contacts possibly creating ephaptic type interactions, and the double synapses themselves being synchronisation agents. The instant the nucleus receives from its afferents a train of excitatory stimuli, these two factors may show themselves to be facilitatory in neuronal synchronisation (Theodosis et al. 1981). It is important to note that the contacts are established exclusively between oxytocin cells, excluding neighbouring vasopressin cells (Theodosis et al. 1986).

Among other mechanisms which might intervene in this synchronisation, the existence of an electrotonic coupling between neurons has been proposed (Andrew et al. 1981). The role of the afferents cannot be excluded, noticeably in explaining the synchronisation between neurons of the different magnocellular nuclei (Belin et al. 1984). These observations should be compared with those of Freund-Mercier and Richard (1984) who showed that oxytocin, by intraventricular injection, facilitated the milk ejection reflex and the electrical activity of oxytocinergic neurons in the lactating rat. Results of in vitro incubation of magnocellular nuclei (Moos et al. 1984) allow one to imagine a local liberation of oxytocin which would thus be capable of exercising a positive feedback action on its own release. Theodosis (1985a) has also described the presence of synapses containing material immunologically reactive to oxytocin in the rat supraoptic nucleus. Intracerebral oxytocin could itself be responsible for neuronal plasticity phenomena accompanying lactation. Preliminary observations show in effect that chronic intraventricular perfusion of oxytocin in the brain of a non lactating rat induces, in eight days, the morphological restructuring of the magnocellular system normally observed during lactation (Theodosis et al. 1986). This is a unique example of induction of anatomical modifications in the adult brain by a neurohormone in order to facilitate its own liberation.

This example of a double central and peripheral action for a neurohormone accompanies, as in the preceding cases, a complimentary behavioural action of the peptide. The injection of oxytocin into the cerebral ventricles of a virgin rat induces, within a few minutes, typical maternal behaviour complete in all aspects: hasty putting together of a nest, regrouping and sheltering of foreign baby rats placed in the cage, licking and recuperation of wandering young (Pedersen *et al.* 1982).

5) Other Neurohormones

Here we shall note simply a certain number of neurohormones whose action takes place in parallel at the periphery and in the central nervous system for cooperation in one and the same regulatory function.

A certain number of GRF's (hypothalamic growth hormone stimulating factors) have been isolated from human pancreatic tumours (Rivier *et al.* 1982, Guillemin *et al.* 1982) and from rat hypothalamus (Spiess *et al.* 1983). They are strongly homologous to glucagon or VIP and thus belong to the vast family of gut related peptides. Their principal actions are exercised by the intermediary of growth hormone on glucose metabolism. Recent observations (Vaccarino *et al.* 1985) show that intraventricular injection of GRF initiates feeding behaviour in the rat, whilst growth hormone itself has no effect.

CCK has been dubbed the satiety hormone (Della-Fera and Baile 1979). This hormone, secreted by the intestine during digestion does not cross the blood brain barrier. It must then be admitted that a CCK is liberated on site in the brain, and would act in parallel with the systemic hormone for the same function (Studler *et al.* 1984).

Luliberin or LH-RH is a hypothalamic hormone which regulates sex gland activity through its action on the hypophyseal gonadotropins. We have already noted that injection of this neurohormone into the preoptic region of the male or female rat induces sexual behaviour within 20–30 minutes (Moss and McCann 1973). According to Riskins and Moss (1979), a nerve pathway using luliberin as messenger could link the hypothalamus to the midbrain, thus coordinating two strategic levels essential in the realisation of the sexual act. It is possible that luliberin takes a neurohormonal route in this communication. The action of luliberin within the nerve structures could be linked to that of

dopamine (Alsatu *et al.* 1981): the actions of these two substances have an autoamplificatory effect in a reciprocal facilitation (Foreman and Moss 1977).

Neurohormonal Space

The concept of a hormone is inseparable from notions of space and duration of action. In its classically accepted form, nervous transmission of information can be regarded as a dialogue between excitable elements in series whereas hormonal information is diffused to an ensemble of target cells dispersed at some distance from the emitting cell. Distance, diffusion and duration of hormonal action are thus in opposition to the discrete and immediate action of conventional neuromediators. We shall consider here the anatomical space in which the neurohormones circulate. We shall take into account at the same time (1) the dispersion of liberation sites other than the classical receiving and secreting poles of nerve cells, (2) the media in which diffusion and circulation of neurohormonal messengers is carried out and finally (3) distribution of the reception sites.

A) Liberation Space

1) Polarisation of Secretory Activity

The existence on the same cell of zones specialised respectively for messenger liberation and reception poses a problem just as much for peptidergic endocrine cells as for certain neurons in the central nervous system.

In the case of endocrine cells, the secretory zone is determined by the neighbouring vessels and connective tissue. In the pancreatic cell, the regions with microvilli, in all probablity receptor sites, are intermingled with smooth regions where hormone liberation preferentially takes place (Fujita *et al.* 1984). In the respiratory tract and the digestive tube, the endocrine cells are frankly polarised, the receptive villous zone faces the lumen whereas the secretory zone is internally orientated.

Among the neurons, a certain number show a net secretory bipolarity. Yuir (1983) observed, in the toad hypothalamus, neurons whose neurosecretory material is accumulated both at a dendritic projection which terminates at the ependymal surface of the 3rd ventricle and on the axonal projection which touches on the external zone of the median eminance and the pericapillary spaces. We shall come back later to the pericapillary problem. An example is provided by the presence of substance P in sensory cell dendrites (Brimijoin *et al.* 1980). The distinction made between dendrites specialised in message reception and conducting axons has no meaning here. According to its pole of liberation, the same neurosecretory material can have different physiological properties. Substance P liberated on cutaneous dendritic terminals acts as a vasodilatatory local hormone, responsible for the axon reflex; the same substance liberated in the medulla serves for sensory message transmission (Lembeck *et al.* 1980). We should remember that this latter hypothesis is the basis of Dale's principle (Dale 1953).

2) Axonal Varicosities and Non-synaptic Liberation

A systematic study of axonal varicosities in the central nervous system has brought morphological arguments to bear on the hypothesis of a non-synaptic liberation of certain chemical messengers (Beaudet and Descarries 1978). This concerns dilatations of 1.2 µm disposed as a string of beads 1 to 3 µm apart, along non myelinated axons (0.1 to 0.2 µm in diameter). These dilatations are filled with pale or dark cored vesicles, mitochondria and all the usual cytological material associated with storage and liberation of a neurotransmitter (Descarries *et al.* 1977). However, the presence of synaptic contacts associated with these varicosities is extremely rare, whether it should concern serotoninergic synapses in the hypothalamus (Calas *et al.* 1974) or the neocortex (Descarries *et al.* 1975), noradrenergic synapses in the neocortex (Descarries *et al.* 1977) or dopaminergic synapses in the neostriatum (Tennyson *et al.* 1974, Arluisson *et al.* 1978).

Autoradiographic definition of aminergic fibres by radioactive amine uptake shows that less than 5% of the varicosities marked by ^{3}H-serotonin and ^{3}H-noradrenaline present synaptic contacts. On the other hand, synaptic junctions are present on more than 50% of non-marked boutons (Descarries *et al.* 1977). For many authors (for review see Vizi 1983), the presence of axonal dilatations is proof of non-synaptic amine liberation in the cortex. These varicosities could form a network throughout the cortical space, contributing to the sprinkling of an "amine flux" over the neurons and the nerve terminals to be found therein.

There is no direct proof of non-synaptic liberation of a chemical messenger in the central nervous system, and although it is suspected, it concerns peptides more than amines. The best demonstrations of non-synaptic messenger liberation are to be found in invertebrate nervous system and the autonomic nervous system of vertebrates.

In vertebrates, the autonomic innervation—intestine, blood vessels—(Burnstock and Costa 1975) provides a credible morphological basis for non-synaptic messenger liberation. A demonstration of non-synaptic liberation is provided by Jan and Jan (1985) concerning a peptide similar to LH-RH in the frog sympathetic ganglion. Stimulation of a C type pre-ganglionic fibre induces a slow excitatory response in a B type post-ganglionic neuron with which there is no synaptic relationship. It has been proved that a peptide similar to LH-RH is responsible for this delayed and prolonged depolarisation (Jan *et al.* 1979). Jan and Jan note that if this type of interneuronal communications favours the diffusion of the message to a large number of cells, it also implies multiplication of the number of messengers to avoid the possibility of interference.

3) An Alternative to Hormonal Diffusion: Tangential Cabling

When a messenger is liberated, its problem lies in reaching the receptor for which it is destined. The specificity of the message derives from the messenger or from its route. As opposed to the peptides, whose number is practically unlimited so that each one can carry one particular message addressed to the target with the specific receptor, the biogenic amines are a restricted group of molecules destined for multiple usage. It is thus important that the amine joins up very precisely with its neuronal target and with no other. The noradrenaline liberated non synaptically in the cortex probably reaches quite uniformly the different neuronal layers and vessels therein. It is desirable on the contrary that only one layer or one neuron type is attained. One can in fact observe a specificity and complementarity concerning the monoaminergic innervation in the different cortical layers. In the rat, the cingulate cortex presents few noradrenergic terminals and these are exclusive to layers V and VI and the deep region of layer I; in parallel, an abundant dopaminergic innervation occupies the superficial part of layer I and layers II and III where afferents of hypothalamic origin terminate (Morrison *et al.* 1979). In these cortical regions the existence of monoaminergic synapses has been clearly demonstrated which coincide, moreover, with the laminar distribution of the afferents. The complementarity of the different monoaminergic innervations implies that each amine reaches precisely the neuronal ensemble that it controls. To draw a parallel with the television information system, this aminergic innervation of the cortex could be compared with telediffusion by cable as opposed to the hormonal modality which is comparable with the hertzien telediffusion. Diffusion of the message by cable is dependant on the extent of the network. Vast cortical surfaces can be reached thanks to the tangential organisation of noradrenergic fibres which run from front to back fanning out over the cortical surface (Morrison et al. 1979). In contrast, the peptidergic innervation of different cortical areas seems to have a radial organisation and is restricted to circumscribed territories in which messenger liberation could take place non-synaptically. A molecular code (specificity of the peptide engaged in a given homeostatic function) would be superposed onto the cable transmission of the neuronal networks (Morrison and Magistretti 1985).

4) Dendritic Liberation

The hypothesis of dendritic transmitter release is relatively recent (Kreutzberg and Toth 1974). Most of the date concerns the substantia nigra (for review see Cheramy et al. 1981).

Björklund and Lindvall (1975) have shown the presence of dopamine in the dendrites of the nigral neurons and postulated the existence of dendro-dendritic synapses. It rapidly appeared that a conventional synaptic junction could hardly account for dopamine activity at the heart of the substantia nigra. Instead of transmitting information from neuron to neuron, the dendritic dopamine could more globally modify information of striatal or pallidal origin arriving in the substantia nigra (Glowinski and Cheramy, 1981). Such action would not require the presence of specialised dendro-dendritic or dendro-axonal synapses (Cuello and Iversen 1978) but would take place by diffusion of the substance in the extracellular space.

Liberation of dopamine in the substantia nigra has been demonstrated in vitro on perifused slices (Geffen et al. 1976) and in vivo by the "push-pull" cannula technique (Glowinski and Cheramy 1981). Evidence of dendritic liberation comes from the following sources: (1) the nigral dendrites are capable of dopamine uptake (Cuello and Iversen 1978); (2) there are no dopaminergic axon terminals or collaterals in the substantia nigra (Wassef et al. 1981); (3) dopamine liberation is greater in the *pars reticulata* which contains almost exclusively dopaminergic dendrites, than in the *pars compacta* containing mostly cell bodies (Cuello and Iversen 1978); (4) although calcium-dependent and provoked by potassium perfusion, dopamine liberation in the substantia nigra is not blocked by tetrodotoxin (Cheramy et al. 1981). As opposed to axonal liberation

which is uniquely concerned with signal transmission, dendritic liberation seems to be responsible for control of afferents. It could intervene at a presynaptic level on afferent gabaergic terminals by facilitating GABA liberation (Reubi et al. 1977). However, most of the proposed mechanisms, implicate the presence of dopaminergic autoreceptors on the dendrites themselves. Such receptors would explain the inhibitory action of dopamine on dopaminergic cells, as illustrated by the effects of local microinfusion of dopamine (Groves et al. 1975).

B) Diffusion Space

In order to reach their respective receptors, the different neurohormonal substances must diffuse in a medium which varies according to their site of liberation: blood, cerebral extracellular space and cerebrospinal fluid. Liberation into the blood takes place via neurohaemal zones: median eminence, vascular organ of the lamina terminalis, neurohypophysis (Calas 1985). We shall limit our account to the two other spaces—cerebrospinal fluid (CSF) and extra-cellular space (ECS)—as they are implicated in the most recently described modes of interneuronal communication.

1) Cerebrospinal Fluid

The idea that the cerebrospinal fluid might act as transport route for neuronal secretion products is relatively

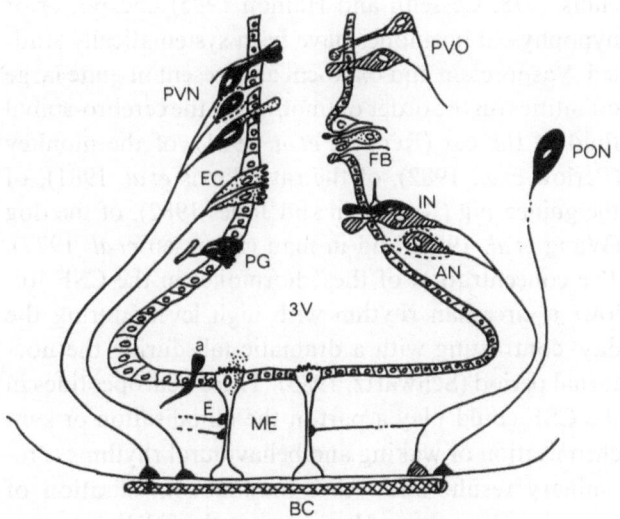

Fig. 7. Schematic representation of the lining of the 3rd ventricle. Abbreviations: *PVN* paraventricular magnocellular neurons; *EC* cyclic ependymal cells; *PG* prechiasmatic gland; *ME* median eminance; *E* ependymal cells of the median eminance; *FB* sensorial formations in "bouquets"; *IN* neurons of the infundibulary nucleus; *AN* neurons of the arcuate nucleus; *a* intraventricular axon

old as it was described under the name of *hydrocriny* by Colin and Barry (1957) as opposed to transport by blood (*hemocriny*) and by neurons (*neurocriny*).

a) Morphological data.

Brightman and Palay (1963) first described in the lining of the 3rd ventricle, a neurosecretory type of nerve terminal. Nerve terminals are also found in the walls of the lateral ventricles (Westergaard 1970) and of the 4th ventricle (Leonhardt and Backhus-Roth 1969). Communication between nerve terminals and ventricular cavities may also be carried out by ependymal cells on which synaptoid contacts can be observed (Rodriguez, 1976). The ependymal cells themselves show secretory activity directed both towards the ventricular spaces and the capillaries of the neurohaemal areas (Fig. 7). We shall leave aside the description and study of the functions of the ependyma and the choroid plexus in order to concentrate on the free passage of CSF to the cerebral extracellular space which contrasts with the remarkable seal between the blood and the CSF unbroken by the choroid plexus or by the exit of the venous valvules (Rowland 1981). The CSF is thus to the cerebral extracellular medium as the plasma is to the systemic fluid medium. It introduces an element of continuity and uniformity in the midst of the neuronal discontinuity and heterogeneity (Vincent 1986).

b) Physiological data.

The presence of numerous neurohormones has been reported in the CSF of which only the endorphins (Terenius 1978, Cesselin and Hamon 1985) and posterior hypophyseal hormones have been systematically studied. Vasopressin and oxytocin are present in quite large quantities (in the order of fmol/ml) in the cerebro-spinal fluid of the cat (Reppert *et al.* 1981), of the monkey (Perlow *et al.* 1982), of the rat (Harris *et al.* 1981), of the guinea-pig (Robinson and Jones 1982), of the dog (Wang *et al.* 1982) and in man (Luerssen *et al.* 1977). The concentration of these hormones in the CSF follows a circadian rhythm with high levels during the day, contrasting with a dramatic fall during the nocturnal period (Schwartz, 1983). These neuropeptides in the CSF could play a part in the initialisation or synchronisation of waking and behavioural rhythms. Preliminary results show that immunoneutralisation of posterior-hypophyseal hormones in the CSF by chronic intraventricular perfusion of antisera leads to a global perturbation of the wake/sleep cycles in the rat (Bibène *et al.* 1985).

Diffusion in the CSF concerns not only posterior-hypophyseal hormones but other numerous neuro-peptides. Dunn (1979) notes that the latter are present in significant concentrations in the CSF and, in addition, the peptidase activity is particularly low. The amines could also employ the CSF as diffusion medium. Chan-Palay (1976) has examined by electron microscopy coupled with autoradiography, the very extended plexus formed by serotoninergic fibres in the ventricular linings. Whilst acknowledging the neurosecretory nature of these terminals this author suggests that they liberate serotonin into the CSF intended for targets at a distance from its site of origin.

2) Cerebral Extracellular Space

The CES represents a diffusion medium for neuronal secretion products at the heart of the central nervous system (Schmitt and Samson 1969, Moller *et al.* 1974, Nicholson 1979). It shows complex architecture subject to permanent restructuring in its glycoprotein frame and in the organisation of the glial cells (Theodosis *et al.* 1985b). The CES represents about 20% of total brain volume. It occupies 18% of the cortex in the adult monkey and no point is further than 15 mm from the CSF (Rapoport 1976).

The diffusion of a substance in the CES has been studied particularly for the case of dendritic dopamine liberation (Glowinski 1981). Dopamine is a relatively small molecule (molecular weight 153). By application of the diffusion formula for a small molecule, in which time is related to the square of the distance, one can estimate that dopamine will diffuse approximately 100 µm in 10 secs, 500 µm in 4 minutes and 1 mm in 17 minutes (Greenfield 1985).

An important point for diffusion of substances in the CES concerns processes of inactivation by recapture or degradation. In general the neurotransmitter molecule once liberated and having acted on its receptor, must be rapidly inactivated in order to interrupt the message. Although reuptake mechanisms can participate in an essential way in this inactivation, as is the case of the monoamines, it is nonetheless true that enzymatic activity on synaptic membranes plays an essential role in the process. The study of enzymatic mechanisms up to now has focussed on the aspect of inactivation by degradation. Certain of these enzymes are well known and have been localised and purified: neutral endopeptidase (EC 24.11) and bestatine sensitive amino-peptidases are typical examples (for reviews see Griffiths and McDermott 1983, Schwartz 1983). However, the concept of specificity of enzyme degradation for a neuropeptide has somewhat weakened. The same is true for a cerebral membrane enzyme activity named early on as enkephalinase (this enzyme detaches the peptide Phe-Met from Met-enkephalin). We know today that it is also capable of degrading substance P (Matsas *et al.* 1983), neurotensin (Checler *et al.* 1983) and neurokinin A (Cooper *et al.* 1985). This substrate specificity opens the possibility for competitive inhibition by this type of enzyme. Thus it has been shown that substance P can inhibit competitively the degradation of cerebral Leu-enkephalin. The analgesic effect of substance P could be due in part to the inhibition of enkephalin degradation (Barclay and Philips 1980).

The notion of neurohormonal space in terms of extracellular metabolism could have important functional significance. This notion is reinforced by recent data showing that extracellular metabolism of neuropeptides, besides its function of inactivation, could continue to refine the intracellular maturation process. In other words, once liberated into the extracellular space, neurohormones could continue to undergo biochemical changes by the action of membrane linked enzymes which could potentiate, inhibit, modify or even invert their biological properties (for review see Griffiths 1983). One of the best examples is the extracellular metabolism of AVP in the brain. Burbach *et al.* (1983) have shown that AVP metabolism by synaptic membranes yields a metabolite (a hexapeptide) more active than AVP itself in memorisation tests, but, on the contrary, devoid of activity on arterial pressure.

This example alone shows that in the case of neurohormones, the molecule secreted by the neuron can be modified during its passage through the extracellular space by enzymes on the external surface of neuronal membranes and so differ from the definitive form acting on the receptor. Thus, the neurohormonal space could very well be presented, not as inert or simply inactivating, but as participating along with intracellular terminal maturation processes in the elaboration of neurohormonal molecules with varying biological potentialities.

C) Reception Area

The action of the messengers liberated into the extracellular space depends on the existence of receptors and on the messenger/receptor interaction. There is therefore a fundamental necessity for reception area, i.e. location of the receptors to be found within the diffusion space of the messengers. If synaptic transmission were the unique means of communication within the central nervous system, the liberation space (pre-synaptic) would be situated precisely with the reception area (post-synaptic). Experience shows that this is not always the case. This "mismatch" between the presence of the messenger in a nerve structure and the anatomical locality of its corresponding receptors argues in favour of neurohormonal communication. The existence of several receptor types for one messenger molecule is another argument in favour of multiplicity of interneuronal communication methods. We shall now consider the problem of mismatching and that of receptor multiplicity.

1) Receptor-messenger Discordance

The remarkable progress of autoradiographic techniques and the development of radioactive analogues have made it possible to establish semi-quantitative receptor maps for different messengers of the central nervous system (Kuhar and Unnerstall 1985). It was logical to compare the distribution of these receptors with the presence of the corresponding messengers in nerve tissue, revealed by radioimmunoassay or immunohistochemistry in brain fragments. However, it appeared that the two did not always coincide. This is the case, for instance, for acetylcholine receptors (Kuhar and Yamamura 1975) for opioid peptides (Simantov *et al.* 1977) for substance P (Quirion *et al.* 1983) for cholescystokinin (Zarbin *et al.* 1983) and for neurotensin (Goedert *et al.* 1984). The case of substance P is particularly spectacular: there is great abundance of immunoreactive product in the substantia nigra (Ljungdal *et al.* 1978) contrasting with an almost total absence of receptors (Quirion *et al.* 1983). A similar discordance is observed in the cerebellum, amygdala and hippocampus in the case of cholecystokinin (Zarbin *et al.* 1983).

An explanation for this discordance is proffered by our outline of the neurohormone concept. As the action takes place at a distance from the liberation site, there is no need for liberation and reception to coincide anatomically. However, the greatest discordance factor, besides anatomical or technical difficulties, remains the multiplicity of receptors for a given messenger. Sometimes a messenger agonist will only fix to a subpopulation of receptors. A characteristic example is that of radioactive N-Methylscopolamine, a muscarinic cholinergic antagonist, which does not recognize the nicotinic receptor. Sometimes it is only a question of affinity which prevents the receptors present in a given structure from recognizing the radioactive ligands at their disposition.

2) Receptor Multiplicity for One Messenger

Substance P is a particularly abundant messenger in the central nervous system (Pernow 1983). As with the opioid peptides, it appears that this peptide is but the main representative of a large family of related molecules, the tachykinins (Quirion 1983). Without going into detail we note the isolation of substance K and neuromedin K, two decapeptides similar to Kassinin and bearing the amino acid sequence characteristic of tachykinins. Moreover Lee *et al.* (1982) have suggested the existence of two subgroups of substance P receptors: the E receptor which mainly recognizes eledoïsine and the P receptor characteristic of physalaemine. Liaison studies using radioactive substance P, substance K and eledoïsine have shown that each ligand has prefered fixation sites in the nervous system and in peripheral tissues. In the substantia nigra are numerous substance K receptors but no substance P or eledoïsine-neuromedin K receptors. The discovery of these different classes of ligands and receptors explains the receptor-messenger mismatch observed for substance P in the substantia nigra as all immunological assays or immunohistochemistry for substance P recognize all related peptides indiscriminately.

Another multiplicity factor concerns the affinity of a receptor for its ligand. The example given by angiotensin II (AII) shows that two sites with differing affinity exist for this peptide on neuronal membranes in the central nervous system (Simonnet and Vincent, 1982) or in primary cultures of spinal cord neurons (Laribi *et al.* 1985): a high affinity site ($K_D = 0.28 \, nM$) and a low affinity site ($K_D = 25 \, nM$). The particular interest of this double affinity lies in its apparent correspondance with two types of electrophysiological action (Legendre *et al.* 1984). When AII is applied by microperfusion pipette in the vicinity of the membrane of a mouse spinal cord neuron in primary culture, two types of electrophysiological response may be observed, depending on the concentration of the AII solution (Fig. 8). At low concentration, equal to or less than $10^{-6} \, M$ in the pipette corresponding to 50 to 100 times less in the medium surrounding the membrane, intracellular recording shows an increase in membrane resistance linked to a decrease in Chloride conductance. At high concentration ($10^{-4} \, M$) one observes on the contrary a decrease in resistance pronounced by an increase in sodium conductance. It must be noted that in physiological terms these two actions, in operation at very different concentrations of the same ligand, could correspond to two modes of messenger action.

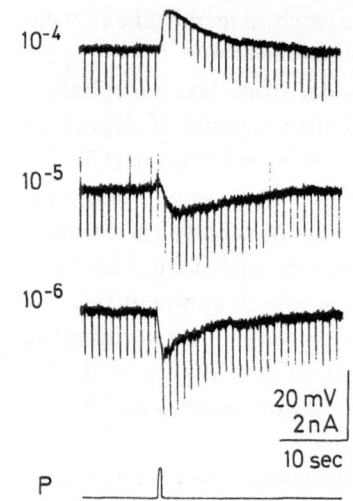

Fig. 8. Electrophysiological response to Angiotensin II (AII) recorded from intracellular electrode in a mouse spinal cord neuron in primary culture. The response differs according to the concentration of A applied near the cell. At a slight concentration of AII ($10^{-6} \, M$), the response consists of a hyperpolarisation with increased membrane resistance (the vertical bars give a mesure of input resistance). At a high concentration ($10^{-4} \, M$) the response is a depolarisation with a decrease in membrane resistance. At $10^{-5} \, M$ an intermediary response is observed

At low concentrations, the messenger could have a neurohormonal type action, the high dilution being due to the large diffusion space. At high concentrations, on the contrary, the messenger would act by the synaptic modality with limited diffusion and so a greater number of molecules in the synaptic cleft. It seems logical that the effect due to a low concentration of the ligand correspond to high affinity receptors while the effect due to a high concentration corresponds to low affinity receptors.

Conclusion

The proof of the messenger lies in the decoding.

In this essay we have tried to show the limits and insufficiencies of the classical opposition between hormones and neurotransmitters. At no point have we evoked the modes of action of these messengers. It is not possible to assume these except by stressing that in this domain too, the difference between hormones and neurotransmitters is difficult to establish. In fact both seem to rely on the intervention of a second messenger in order to establish the stimulus-secretion coupling. When we have put aside the case of direct coupling between a receptor and a voltage independent ionic channel which corresponds to the classical understanding of the neurotransmitter (Barker *et al.*

1980), there remain cases, apparently numerous, where the coupling is carried out via a second messenger. In this domain the nervous systems that have been studied do not show much originality in comparison with hormonal systems (Miller 1985).

Free intracellular calcium constitutes the common agent of all regulatory mechanisms employing calcium dependant enzymes. The concentration of free Ca^{2+} within the cell depends either on the intervention of voltage dependant membrane channels, or on the mobilisation of internal reserves. These two mechanisms can be observed for one messenger. In the sensory neurons of Aplysia, serotonin facilitates calcium entry by blocking a potassium channel which normally holds the membrane in a hyperpolarised state (Siegelbaum *et al.* 1982). But at the neuromuscular junction of the crayfish, Kravitz *et al.* (1985) have shown that serotonin facilitates liberation of the mediator by mobilisation of internal calcium reserves. Calcium is also the indispensable agent for exocytosis of the vesicle or granule containing the secretory product. Here again there is hardly any difference between hormone and neurotransmitter. All the same it seems that a certain neuronal specificity exists concerning the protein implicated in calcium intervention: synapsin I (Bruwning *et al.* 1985). It is, effectively, only present in neurons; it is absent, for instance, from chromaffin cells. Calcium dependent phosphorylation of this protein apparently liberates the vesicle from its cyto-skeletal attachments thus allowing its release (Miller 1985). It remains to be seen whether neurons of the central nervous system implicated in the neurohormonal transmission process contain synapsin I or not. This would furnish a veritable demarcation line between hormonal transmission and neuronal transmission.

Acknowledgements

We heartily thank our colleagues, L. Périer, E. Arnauld, and F. Rodriguez, for their help with the production of this manuscript.

References

Akaishi T, Negoro H, Kobayashi S (1980) Responses of paraventricular and supraoptic units to angiotensin II, [Sar^{1}Ile^8]-angiotensin II and hypertonic NaCl administered into the cerebral ventricle. Brain Res 188: 499–511

Alsatu M, Kempf E, Mack G, Aron C (1981) Involvement of dopaminergic mechanisms in the control of ovulation and sexual receptivity in cyclic female rats. Biol Behav 6: 305–315

Andrew RD, MacVicar BA, Dudek FE, Hatton GI (1981) Dye transfer through gap junctions between neuroendocrine cells of rat hypothalamus. Science 211: 1187–1189

Antunes-Rodrigues J, McCann SM, Rogers LC, Samson WK (1986) A trial natriuretic factor inhibits dehydration- and angiotensin II-induced water intake in the conscious, unrestrained rat. Proc Natl Acad Sci 82: 8720–8723

Arluisson M, Agid Y, Javoy F (1978) Dopaminergic nerve endings in the neostriatum of the rat. I. Identification by intracerebral injection of 6 hydroxydopamine. Neuroscience 3: 657–673

Barclay RK, Philipps MA (1980) Inhibition of enkephalin-degrading aminopeptidase activity by certain peptides. Biochem Biophys Res Commun 96, 4: 1732–1738

Barker JL, Gruol DL, Huang LHM, MacDonald JF, Smith Jr TG (1980) Electrophysiological analysis of the role of peptides using cultured spinal neurons in the role of peptides in neuronal function, Barker JL (ed). Marcel Dekker Inc., New York, pp 273–300

Barnard RR, Morris M (1982) Cerebro-spinal fluid vasopressin and oxytocin: evidence for an osmotic response. Neurosci Lett 29: 275–279

Beaudet A, Descarries L (1978) The monoamine innervation of rat cerebral cortex; synaptic and non-synaptic axon terminals. Neuroscience 3: 851–860

Belin V, Moos F, Richard P (1984) Synchronization of oxytocin cells in the hypothalamic paraventricular and supraoptic nuclei in suckled rats: direct proof with paired extracellular recordings. Expl Brain Res 5: 201–203

Berson SA, Yalow RS (1973) Peptides hormones. Part II. In: Pituitary hormones and hypothalamic releasing factors. Elsevier North-Holland, Amsterdam, pp 257–711

Biales B, Dichter MS, Tischler A (1976) Electrical excitability of cultured adrenal chromaffin cells. J Physiol (London) 262: 743–753

Bibène V, Pestre M, Rodriguez F, Arnauld E, Poncet C, Vincent JD (1985) Vasopressine et cycle veille-sommeil chez le rat. 15e Coll Soc Neuroendocrinol Exptl, Gif-sur-Yvette

Björklund A, Lindvall D (1975) Dopamine in dendrites of substantia nigra neurons: suggestions for a role in dendritic terminals. Brain Res 83: 531–537

Brightman MW, Palay SL (1963) The fine structure of ependyma in the brain of the rat. J Cell Biol 19: 415–439

Brimijoin S, Lundberg JM, Brodin E, Hökfelt T (1980) Axonal transport of substance P in the vagus and sciatic nerves of the guinea-pig. Brain Res 191: 443–457

Bruwning MD, Huganir R, Greengard P (1985) Protein phosphorylation and neuronal function. J Neurochem 45: 11–23

Buijs RM, Swaab DF, Dogterom J, Van Leeuwen FW (1978) Intra- and extrahypothalamic vasopressin and oxytocin pathways in the rat. Cell Tissue Res 186: 423–483

Burbach JPH, Kovacs GL, De Wied D, Van Nispen JW, Greven HM (1983) A major metabolite of arginine vasopressin in the brain is a highly potent neuropeptide. Science 221: 1310–1312

Burnstock G, Costa M (1975) Adrenergic neurons: their organisation, function and development in the peripheral nervous system. Chapman et Hall, London

Calas A (1985) Morphological correlates of chemically specified neuronal interaction in the hypothalamo-hypophyseal area. Neurochem Int 7: 921–940

— Alonso G, Arnauld E, Vincent JD (1974) Demonstration of indolaminergic fibers in the median eminence of the duck, rat and monkey. Nature (Lond) 250: 241–243

Cesselin F, Hamon M (1985) Significations fonctionelles possibles de la libération simultanée de plusieurs neurotransmetteurs putatifs par un même neurone. Annales d'endocrinologie 45: 207–213

Chan-Palay V (1976) Serotonin axons in the supra and subependymal plexuses and in the leptomeninger; their roles in local alteration of cerebrospinal fluid and vaso motor activity. Brain Res 102: 103–130

Checler F, Vincent JP, Kitabi P (1983) Degradation of neurotensin by rat brain synaptic membranes; involvement of a thermolysin-like metallo-endopeptidase (enkephalinase), angiotensin-converting enzyme, and other unidentified peptidase. J Neurochem 41: 375–384

Cheramy A, Leviel V, Glowinski J (1981) Dendritic release of dopamine in the substancia nigra. Nature (Lond) 289: 537–542

Cocchia D, Miani N (1980) Immunocytochemical localization of the brain specific S-10 protein in the pituitary gland of adult rat. J Neurocytol 9: 771–782

Colin R, Barry J (1957) Neurosécrétion et diabète insipide. Histophysiologie de la neurosécrétion. Ann Endocrinol 18: 464–469

Cooper NM, Kenny AJ, Turner AJ (1985) The metabolism of neuropeptides—Neurokinin A (substance K) is a substrate for endopeptidase—24.11 but not for peptidyl dipeptidase A (angiotensinconverting enzyme). Biochem J 231: 357–361

Cuello AC, Iversen LL (1978) Interactions of dopamine with others neurotransmitters in the rat substantia nigra: a possible functional role of dendritic dopamine. In: Garattini S, Pujol JF, Samanin R (eds) Interactions between putative neurotransmitters in the brain. Raven Press, New York, pp 127–150

Dale HH (1935) Pharmacology and nerve endings. Proceedings of the Royal Society of Medicine 28: 319–332

— (1953) Adventures in physiology. Pergamon Press, London

Della-Fera MA, Baile CA (1979) Cholecystokinin octapeptide: continuous picomole injections into the cerebral ventricules of sheep suppress feeding. Science 206: 471–473

Demotes-Mainard J, Chauveau J, Rodriguez F, Vincent JD, Poulain DA (1986) Septal release of vasopressin in response to osmotic, hypovolemic and electrical stimulation in rats. Brain Res (in press)

Descarries L, Beaudet A, Watkins KC (1975) Serontonin nerve terminals in adult rat neo-cortex. Brain Res 100: 563–588

— Beaudet A (1983) The use of radio-autography for investigating transmitter-specific neurons. In: Björklund A, Hökfelt T (eds) Handbook of chemical neuroanatomy. Elsevier, Amsterdam, pp 286–364

— Watkins KC, Lapierre Y (1977) Noradrenergic axon terminals in the cerebral cortex of rat. III Topometric ultrastructural analysis. Brain Res 133: 197–222

De Wied D (1965) The influence of the posterior and intermediate lobe of the pituitary and pituitary peptides on the maintenance of a conditioned avoidance response in rats. Int J Neuropharmacol 4: 157–167

Dismukes RK (1979) New concepts of molecular communication among neurons. The Behavioral and Brain Sciences 2: 409–448

Doris PA, Bell FR (1984) Vasopressin in plasma and cerebrospinal fluid of hydrated and dehydrated steers. Neuroendocrinol 38: 290–296

Dun NJ, Nishi S, Karczman AG (1978) An analysis of the effect of angiotensin II on mammalian ganglion cells. J Pharmacol exp Ther 204: 669–675

Dunn AJ (1979) Molecular signals released by neurons. The Behavioral and Brain Sciences 2: 422–423

Emson PC (1985) Neurotransmitter systems. In: Bousfield D (ed) Neurotransmitter in action. Elsevier Biomedical Press, Amsterdam, pp 6–10

Epstein Y, Castel M, Glick SM, Sivan N, Ravid R (1983) Changes in hypothalamic and extra-hypothalamic vasopressin content of water-deprived rats. Cell Tiss Res 233: 99–111

Felix D, Akert K (1974) The effect of angiotensin II on neurons of the cat subfornical organ. Brain Res 76: 350–353

Fitzsimons JT (1972) Thirst. Physiol Rev 52: 468–559

Foreman MM, Moss RL (1977) Effects of subcutaneous injection and intrahypothalamic infusion of releasing hormones upon lordotic response to repetitive coïtal stimulation. Horm Behav 8: 219–234

Freund-Mercier MJ, Richard P (1984) Electrophysiological evidence for facilitatory control of oxytocin neurones by oxytocin during suckling in the rat. J Physiol (Lond) 352: 447–466

Fujita T (1977) Concept of paraneurones. In: Kobayashi S, Chiba T (eds) Paraneurones: new concepts on neuroendocrine relatives. Japan Soc Histol Documentation, Niigata, pp 1–12

— Kobayashi S, Uchida T (1984) Secretory aspect of neurons and paraneurons. Biochemical Res [suppl] 5: 1–8

Fuxe K, Ganten D, Hökfelt T, Bolme P (1976) Immunohistochemical evidence for the existence of angiotensin II containing nerve terminals in the brain and spinal cord of the rat. Neurosci Lett 2: 229–234

Ganten D, Fuxe K, Phillips MI, Mann JFE, Ganten U (1978) The brain isorenin-angiotensin system: histochemistry localization and possible role in drinking and blood pressure regulation. In: Ganong WF, Martini B (eds) Frontiers in neuroendocrinology, Vol. 5. Raven Press, New York, pp 61–99

Geffen LB, Jessel TM, Cuello AC, Iversen LL (1976) Release of dopamine from dendrites in rat substantia nigra. Nature (Lond) 260: 258–260

Glowinski J, Cheramy A (1981) Dentritic release of dopamine its role in the substantia nigra. In: Stjarne L, Hedquist P, Lagercranz H, Wennmalm A (eds) Chemical transmission: 75 years. Academic Press, New York, pp 285–299

Goedert M, Mantyh PW, Emson PC, Hunt SP (1984) Inverse relationship between neurotensin receptors and neurotensin-like immunoreactivity in cat striatum. Nature (Lond) 307: 543–546

Greenfield SA (1985) The significance of dendritic release of transmitter and protein in the substantia nigra. Neurochem Int 7: 887–901

Griffiths EC, MacDermott JR (1983c) Biotransformation of neuropeptides. Progress in Neuroendocrinology 39: 573–581

Gronan RJ, York DH (1978) Effects of angiotensin II and acetylcholine on neurones in the preoptic area. Brain Res 154: 172–177

Groves PM, Wilson CJ, Young SJ, Rebec GU (1975) Self-inhibition by dopaminergic neurones. Science 190: 522–529

Guillemin R, Brazeau P, Böhlen P, Esch F, Ling N, Wehrenberg WB (1982) Growth hormone-releasing factor from a human pancreatic tumor that caused acromegaly. Science 218: 585–587

Harris MC, Jones PM, Robinson ICAF (1981) Differences in the release of oxytocin into blood and cerebrospinal fluid following hypothalamic and pituitary stimultion in rats. J Physiol (Lond) 320: 109–110P

Hill C, Hendry IA (1977) Development of neurones synthetizing noradrenaline and acetylcholine in the superior cervical ganglion of the rat in vivo and in vitro. Neurosci 2: 741–749

Hökfelt T, Johanson O, Goldstein M (1984) Chemical anatomy of the brain. Science 225: 1326–1334

Hughes J, Smith TW, Kosterlitz HW, Fothergill TH, Morgan BA, Morris HR (1975) Identification of two related pentapeptides from the brain with potent opiate agonist activity. Nature (Lond) 258: 577–580

Huwyler T, Felix D (1980) Angiotensin II-sensitive neurons in septal areas of the rat. Brain Res 195: 187–195

Iovino M, Poenaru S, Annunziato L (1983) Basal and thirst-evoked vasopressin secretion in rats with electrolytic lesion of the medioventral septal area. Brain Res 258: 123–126

Jan YN, Jan LY (1985) A LH-RH-like peptidergic neurotransmitter capable of "acton at a distance" in autonomic ganglion. In: Bousfield D (ed) Neuro-transmitters in action. Elsevier Biomedical Press, Amsterdam/New York/Oxford, pp 94–103

— — Kuffler SW (1979) A peptide as a possible transmitter in sympathetic ganglia of the frog. Proc Natl Acad Sci USA 76: 1501–1505

Joels M, Urban IJA (1982) The effect of microiontophoretically applied vasopressin and oxytocin on single neurons in the septum and dorsal hippocampus of the rat. Neurosci Lett 33: 79–84

Knight DP (1970) Sclerotisation of the perisarc of the calyptoblastic hydroid laomedea fluxuosa. Tissue and Cell 2: 467–477

Kobayashi S (1977) Adrenal medulla: chromaffin cells as paraneurones. Arch Histol Jap 40: 61–70

Kravitz EA, Beltz BS, Glusman S, Goy MF, Harris-Warrick RM, Johnston MF, Livingstone MS, Schwarz TL, Siwicki KK (1985) Neurohormones and lobsters: biochemistry to behavior. In: Bousfield D (ed) Neurotransmitters in action. Elsevier Biomedical Press, Amsterdam/New York/Oxford, pp 135–142

Kreutzberg GW, Toth L (1974) Dendritic secretion: a way for the neuron to communicate with the vasculature. Naturwissenschaften 61: 37–39

Krieger DT (1983) Brain Peptides: what, where and why? Science 222: 975–985

Kuhar MJ, Unnerstall JR (1985) Quantitative receptor mapping by autoradiography: some current technical problems. TINS 8: 49–53

— Yamamura HI (1975) Light autoradiographic localisation of cholinergic muscarinic receptors in rat brain by specific binding of a potent antagonist. Nature (Lond) 253: 560–561

Laribi C, Legendre P, Dupouy B, Vincent JD, Simonnet G (1985) Characterization of two angiotensin II binding sites in cultured mouse spinal cord neurones. Brain Res 347: 94–103

Larsson LI (1980) On the possible existence of multiple endocrine, paracrine and neurocrine messengers in secretory cell systems. Invest Cell Pathol 3: 73–85

Le Douarin N (1982) The neural crest. In: Barlow PW, Green PB, Wylie CC (eds) Developmental and cell biology series. Cambridge University Press, Cambridge

— Teillet MA (1973) The migration of neural crest cells to the wall of the digestive tract in avian embryo. J Embryol Exp Morphol 30: 31–48

Lee CM, Iversen LL, Hanley MR, Sandberg BEB (1982) The possible existence of multiple receptors for substance P. Naunyn-Schmiedebergs Arch Pharmacol 318: 281–287

Legendre P, Simonnet G, Vincent JD (1984) Electrophysiological effects of angiotensin II on cultured mouse spinal cord neurones. Bain Res 297: 287–296

Lembeck F, Gamse R, Holzer P, Molnar A (1980) Substance P and chemosensitive neurones. In: Ajmone-Marsan C, Traczyk WZ (eds) Neuropeptides and neural transmission. Raven Press, New York, pp 51–72

Le Moal M, Koob GF, Koda LY, Bloom FE, Manning M, Sawyer

WH, Rivier J (1981) Vasopressin receptor antagonist prevents. Nature (Lond) 291: 491–493

Leonhardt H, Backhus-Roth A (1969) Synapsenartige Kontakte zwischen intraventrikulären Axonendigungen und freien Oberflächen von ependymzellen des Kaninchengehirns. Zellforschung und mikroskopischen Anatomie 97: 369–376

Le Roith D, Liotta AS, Roth J, Schiloach J, Lewis ME, Pert CB, Krieger DT (1982) Corticotrophin and β-endorphin-like materials are native to unicellular organisms (tetrahymena). Proc Natl Acad Sci USA 79: 2086–2090

Liotta AS, Loudes C, McKelvy JF, Krieger DT (180) Biosynthesis of the precursor corticotrophin/endorphin-, corticotropin-, melanotropin-β lipotropin-, and β-endorphin-like material by cultured neonatal rat hypothalamic neurons. Proc Natl Acad Sci USA 77: 1880–1884

Ljungdal A, Hökfelt T, Nilsson G (1978) Distribution of substance P-like immunoreactivity in the central nervous system of the rat. I. Cell Neuroscience, 3: 861–943

Llinas R, Greenfield SA, Jahnsen H (1984) Electrophysiology of pars compacta cells in the in vitro substancia nigra—a possible mechanism for dendritic release. Brain Res 294: 127–132

Loumaye E, Thorner J, Catt KJ (1982) Yeast mating pheromone activates mammalian gonadotrophs: evolutionary conservation of a reproductive hormone? Science 218: 1323–1325

Lubar JF, Boyce BA, Schaeffer CF (1968) Etiology of polydipsia and polyuria in rats with septal lesions. Physiol Behav 3: 289–292

Luerssen TG, Shelton RL, Robertson GL (1977) Evidence for separate origin of plasma and cerebrospinal fluid vasopressin. Clin Res 25: 14A

Lundberg JM (1981) Evidence for co-existence of vasoactive intestinal polypeptide (VIP) and acetylcholine in neurones of cat exocrine glands. Morphological, Biochemical and Functional Studies. Acta Physiol Scand [suppl] 496: 1–57

Matsas R, Fulcher IS, Kenny AJ, Turner AJ (1983) Substance P and [Leu] enkephalin are hydrolyzed by an enzyme in pig caudate synaptic membranes that is identical with the endopeptidase of kidney microvilli. Proc Natl Acad Sci USA 80: 3111–3115

Matthews EK, Sakamoto Y (1975a) Electrical characteristics of pancreatic islet cells. J Physiol (Lond) 246: 421–437

— — (1975b) Pancreatic islet cells: electrogenic and electrodiffusional control of membrane potential. J Physiol (Lond) 246: 439–457

Miller RJ (1985) Second messengers, phosphorylation and neurotransmitter release. TINS 8: 463–465

Moller M, Mollergärd K, Lund-Andersen H, Hertz L (1974) Concordance between morphological and biochemical estimates of fluid spaces in rat brain cortex slices. Expl Brain Res 21: 299–314

Moos F, Freund-Mercier MJ, Guerne Y, Guerne JM, Stueckel ME, Richard P (1984) Release of oxytocin and vasopressin by magnocellular nuclei in vitro: specific facilitatory effect of oxytocin on its own release. J Endocrinol 102: 63–72

Morris JF, Nordmann JJ, Dyball REJ (1978) Structure-function correlation in mammalian neurosecretion. Int Rev Exp Path 18: 1–95

Morrison JH, Magistretti PJ (1985) Monoamines and peptides in cerebral cortex. Contrasting principles of cortical organization. In: Bousfield D (ed) Neurotransmitter in action. Elsevier Biomedical Press, Amsterdam/New York/Oxford, pp 319–328

Morrison JH, Molliver ME, Grzanna R (1979) Noradrenergic innervation of cerebral cortex: widespread effects of local cortical lesions. Science 205: 313–316

Moss RL, McCann SM (1973) Induction of making behavior in rats by luteinizing hormone-releasing factor. Science 181: 177–179

Nakajima T, Yamaguchi H, Takahashi K (1980) S-100 protein in folliculostellate cells of rat pituitary anterior lobe. Brain Res 191: 523–531

Nicholson C (1979) Brain cell microenvironment as a communicative channel. In: Schmitt FO, Worden FG (eds) The neurosciences fourth study programm. MIT Press, Cambridge, Massachusetts, pp 457–476

Pearse AGE (1966a) 5-Hydroxytryptophan uptake by dog thyroid C cells and its possible significance in polypeptide hormone production. Nature (Lond) 211: 598–600

— (1969) The cytochemistry and ultastructure of polypeptide hormone-producing cells of the APUD series and the embryologie, physiologie and pathologie implication of the concept. J Histochem Cytochem 17: 303–313

— (1983) The neuroendocrine division of the nervous system: APUD cells as neurones or paraneurones. In: Osborne NN (ed) Dale's principle and communication between neurones. Pergamon Press, Oxford, pp 37–48

Pedersen CA, Ascher JA, Monroe YL, Prange (Jr) (1982) Oxytocin induces maternal behavior in virgin female rats. Science 216: 648–650

Pelletier G, Steinbusch HWM, Verhufstad AAJ (1981) Immunoreactive substance P and serotonin present in the same dense-core vesicles. Nature (Lond) 293: 71–72

Perlow MJ, Reppert SM, Artman HA, Fisher DA, Seif SM, Robinson AG (1982) Oxytocin, vasopressin, and estrogen-stimulated neurophysin: daily patterns of concentrations in cerebrospinal fluid. Science 216: 1416–1418

Pernow B (1983) Substance P. Pharmacol Rev 35: 85–141

Pestre M, Arnauld E, Vincent JD (1984) Actions of micro- iontophoretically applied vasopressin selective agonists and antagonists on single neurons in the lateral septum of the rat. Neurosci Lett [suppl] S 341

Poulain DA, Ellendorff F, Vincent JD (1980) Septal connections with identified oxytocin and vasopressin neurones in the supraoptic nucleus of the rat. An electrophysiological investigation. Neurosci 5: 379–387

— Lebrun CJ, Vincent JD (1981) Electrophysiological evidence for connections between septal neurones and the supraoptic nucleus of the hypothalamus of the rat. Exp Brain Res 42: 260–268

— Wakerley JB (1982) Electrophysiology of hypothalamic magnocellular neurones secreting oxytocin and vasopressin. Neuroscience 7: 773–808

Quirion R, Shults CW, Moody TW, Pert CB, Chase TN, O'Donohue TL (1983) Autoradiographic distribution of substances P receptors in rat central nervous system. Nature (Lond) 303: 714–716

Rapoport SI (1976) Blood-brain barrier in physiology and medicine. Raven Press, New York

Reppert SM, Artman HG, Swaminathan S, Fisher DA (1981) Vasopressin exhibits a rhythmic daily pattern in cerebro spinal fluid but not in blood. Science 213: 1256–1257

Reubi JC, Iversen LL, Jessel TM (1977) Dopamine selectively increases [3] GABA release from slices of rat substantia nigra in vitro. Nature (Lond) 268: 652–654

Riskins P, Moss RL (1979) Midbrain central gray: LH-RH infusion enhances lordotic behavior in oestrogen primed ovariectomized rats. Brain Res Bull 4: 203–205

Rivier C, Vale W (1983) Interaction of corticotropin-releasing factor and arginine vasopressin on adrenocorticotropin in vivo. Endocrinology 113: 939–942

Robinson IFA, Jones PM (1982) Neurohypophyseal peptides in cerebrospinal fluid: recent studies. In: Baertschi AJ, Dreifuss JJ (eds) Neuroendocrinology of vasopressin, corticoliberin and opiomelanocortins. Academic Press, New York, pp 21–31

Rodriguez EM (1976) The cerebrospinal fluid as a pathway in neuroendocrine integration. J Endocrinol 71: 407–443

Rodriguez F, Demotes-Mainard J, Chauveau J, Poulain DA, Vincent JD (1983) Vasopressin release in the rat septum in response to systemic stimuli. Soc Neurosci Abstr 9: 445

Rowland LP (1981) Blood-brain barrier, cerebrospinal fluid, brain edema and hydrocephalus. In: Kandel ER, Schwartz JH (eds) Principles of neural science. Elsevier-North Holland, New York/Amsterdam/Oxford, pp 651–659

Samson WK (1985) Atrial natriuretic factor inhibits dehydration and hemorrhage-induced vasopressin release. Neuroendocrinology 40: 277–279

Saper CB, Standaert DG, Currie MG, Schwartz D, Geller DM, Needleman P (1985) Atriopeptin-immunoreactive neurons in the brain: presence in cardiovascular regulatory areas. Science 227: 1047–1049

Schmitt FO, Samson FE (1969) Brain cell microenvironment. Neurosci Res Prog Bull 7: 277–417

Schwabe C, Le Roith D, Thompson RP, Shiloach J, Roth J (1983) Relaxin extracted from protozoa (Tetrahymena lyriformis). Molecular and immunologic properties. J Biol Chem 258: 2778–2781

Schwartz JC (1983) Metabolism of enkephalins and the inactivating a neuropeptidase concept. Trends Neurosci 6: 45–48

Schwartz D, Geller DM, Manning PT, Siegel NR, Fok KF, Smith CE, Needleman P (1985) Ser-Leu-Arg-Atriopeptin III: the major circulating form of atrial peptide. Science 229: 397–400

Sherrington CS (1906) The integrative action of the nervous system. Yale University Press, New Haven

Shibuki K (1984) Supraoptic cells: synaptic inputs from the nucleus accumbens in the rat. Exp Brain Res 53: 341–348

Sibole W, Miller JJ, Mogenson GJ (1971) Effects of septal stimulations on drinking elicited by electrical stimulation of the lateral hypothalamus. Exp Neurol 32: 466–477

Siegelbaum SA, Camardo JS, Kandel ER (1982) Serotonin and cyclic AMP close single, K^+ channels in Aplysia sensory neurones. Nature (Lond) 299: 413–417

Simantov R, Kuhar MJ, Uhl GR, Snyder SH (1977) Opioid peptide enkephalin: immunohistochemical mapping in rat central nervous system. Proc Natl Acad Sci. USA 74: 2167–2175

Simonnet G, Bioulac B, Rodriguez F, Vincent JD (1980) Evidence for a direct action of angiotensin II on neurones in the septum and in the medial preoptic area. Pharmacol Biochem Behav 13: 359–363

— Carayon A, Allard M, Cesselin F, Lagoguey A (1984) Evidence for an angiotensin II-like material and for a rapid metabolism of angiotensin II in the rat brain. Brain Res 304: 93–103

— Rodriguez F, Fumoux F, Czernichow P, Vincent JD (1979) Vasopressin release and drinking induced by intracranial injection of angiotensin II in monkey. Am J Physiol 237: R20–R25

— Vincent JD (1982) Characteristics of angiotensin II binding sites in the neostriatum of the rat brain. Neurochem Int 4: 149–155

Sofroniew MV (1985) Vasopressin and oxytocin in the mammalian brain and spinal cord. In: Bousfield D (ed) Neurotransmitters in action. Elsevier Biomedical Press, Amsterdam/New York/Oxford, pp 329–337

Spiess J, Rivier J, Vale W (1983) Characterization of a rat hypothalamic growth hormone-releasing factor. Nature 303: 532–535

Suga T, Suzuki M (1979) Effects of angiotensin II on the medullary neurons and their sensitivity to acetylcholine and catecholamines. Jap J Pharmacol 29: 541–552

Studler JM, Simon M, Cesselin F, Blanc G, Glowinski J, Tassin JP (1984) Pharmacological study on the mixed CCK 8/DA meso nucleus accumbens pathway: evidence for the existence of vesicles, containing the two transmitters. Brain Res 298: 91–97

Standaert DG, Saper CB, Needleman P (1985) Atriopeptin: potent hormone and potential neuromediator. TINS 8: 510–511

Szczepanska-Sadowska E, Gray D, Simon-Opperman C (1983) Vasopressin in blood and third ventricle CSF during dehydration, thirst and hemorrhage. Am J Physiol 245: R549–R555

Takor-Takor T, Pearse AGE (1975) Neuroectodermal origin of avian hypothalamo-hypophyseal complex: the role of the ventral neural midge. J Embryol Exp Morphol 34: 311–325

Tanaka I, Misono KS, Inagami T (1984) Atrial natriuretic factor in rat hypothalamus, atria and plasma: determination by specific radioimmunoassay. Biochem Biophys Res Commun 124: 663–668

Teitelman G, Joh TH, Reis DJ (1981) Transformation of catecholaminergic precursors intoglucagon (A) cells in mouse embryonic pancreas. Proc Natl Acad Sci USA 78: 5225–5229

Tennyson VM, Heikkila R, Mytilineau C, Cote L, Cohen G (1974) 5-Hydroxydopamine "tagged" neuronal boutons in rabbit neostriatum: interrelationship between vesicles and axonal membrane. Brain Res 82: 341–348

Terenius L (1978) Endogenous peptides and analgesia. Ann Rev Pharmacol Toxicol 18: 189–204

Theodosis DT, Poulain DA, Vincent JD (1981) Possible morphological bases for synchronisation of neuronal firing in the rat supraoptic nucleus during lactation. Neuroscience 6: 919–929

— (1985a) Oxytocin-immunoreactive terminals synapse on oxytocin neurones in the supraoptic nucleus. Nature (Lond) 313: 682–684

— Chapman DB, Montagnese C, Poulain DA, Morris JF (1985b) Structural plasticity in the hypothalamic supraoptic nucleus at lactation affects oxytocin but not vasopressin secreting neurones. Neuroscience, (in press)

— Montagnese C, Rodriguez F, Vincent JD, Poulain DA (1986) Oxytocin induces morphological plasticity in the adult hypothalamo-neurohypophysial system. Nature (Lond) 322: 738–740

Tsong SD, Philipps D, Halmi N, Liotta AS, Margioris A, Bardin CW, Krieger DT (1982) ACTH and beta-endorphin related peptides are present in multiple sites in the reproductive tract of the male rat. Endocrinology 110: 2204–2206

Vaccarino FJ, Bloom FE, Rivier J, Vale W, Koob GF (1985) Stimulation of food intake by centrally administered hypothalamic growth hormone-releasing factor. Nature (Lond) 314: 167–168

Vanderhaeghen JJ, Signeau JC, Gepts W (1975) New peptide in the vertebrate CNS reacting with antigastrin antibodies. Nature (Lond) 257: 604–605

Van Wimersma Greidanus TB (1982) Disturbed behavior and memory of the Brattleboro rat. Ann NY Acad Sci 394: 655–662

Vincent JD, Israel JM, Brigant JL (1985) Ionic channels in hormone release from adenohypophysial cells—an electrophysiological approach. Neurochem Int 7: 1007–1016

Vincent JD (1986) Biologie des passions. Odile Jacob—Le Seuil, Paris

Vizi ES (1983) Non synaptic interneuronal communication: Physiological and pharmacological implication. In: Osborne N (ed) Dale's principle and communication between neurones. Pergamon Press, Oxford, pp 83–111

Wang BC, Share L, Crofton JT, Kimura T (1982) Effects of intravenous and intracerebroventricular infusion of hypertonic solutions on plasma and cerebrospinal fluid vasopressin concentrations. Neuroendocrinology 34: 215–221

Wassef M, Berod A, Sotelo C (1981) Dopaminergic dendrites in the pars reticulata of the rat substantia nigra and their striatal input-combined immunocytochemical localization of tyrosine hydroxylase and anterograde degeneration. Neuroscience 6: 2125–2139

Wayner MJ, Ono T, Nolley D (1973) Effect of angiotensin II on central neurons. Pharmacol Biochem Behav 1: 679–691

Westergaard E (1970) The lateral ventricles and the ventricular walls. Thesis. Arhus, Danemark

Yuir (1983) Immunohistochemical studies on peptide neurons in the hypothalamus of the bullfrog Rana Catesbliana. Gen Comp Endocrinol 49: 195–209

Zarbin MA, Innis RB, Wamsley JK, Snyder SH, Kuhar MJ (1983) Autoradiographic localization of cholecystokinin receptors in rodent brain. J Neuroscience 4: 877–906

Zerbe RL, Palkovits M (1984) Changes in the vasopressin content of discrete brain regions in response to stimuli for vasopressin secretion. Neuroendocrinology 38: 285–289

Correspondence: J. D. Vincent, M.D., INSERM U. 176, Domaine de Carreire, Rue Camille Saint-Saëns, F-33077 Bordeaux Cédex, France.

Acta Neurochirurgica, Suppl. 47, 38–41 (1990)

Central Control of Circadian and Ultradian Neuroendocrine Rhythms

I. Assenmacher

Laboratory Endocrinological Neurobiology, UA 1197-CNRS, University of Montpellier II, Montpellier, France

The very first pioneers of the concept of homeostasis, particularly Henderson, stated as long ago as the twenties that within the homeostatic range the fluctuations of physiological parameters follow a rhythmic pattern and such biological rhythms play an adaptive role. A deeper insight into the mechanisms underlying biological rhythms has only been gained in the past two decades (Krieger 1979, Aschoff 1981, Moore-Edde *et al.* 1982). The most ubiquitous biological oscillators appeared to be circadian rhythms, that is, rhythms with a 24 h periodicity. These were evident not only over the whole zoological scale up to humans, but also at all levels of the organism, from the basic cellular processes up to the most integrated functions. Moreover, recent studies in molecular biology have provided evidence that the circadian pattern of biological rhythms has a genetic basis (Bargiello *et al.* 1984, Reddy *et al.* 1984).

Generally speaking, two further properties of circadian rhythms are of physiological importance. Firstly, there are fixed temporal relationships between the basic behavioral and neuroendocrine circadian rhythms, which persist in the free-running state (Ixart *et al.* 1983b). For instance, in all organisms explored so far, the ACTH-corticosteroid rhythm starts its ascending phase during the second half of the sleep phase, so that humans and animals start their diurnal activity phase with maximal production of glucocorticosteroids and consequently an adequate blood sugar availability, whatever the feeding possibilities may be. Secondly, there are close relationships between the circadian oscillation of any particular function and its shorter-lived ultradian fluctuations, the latter appearing earlier in the course of ontogeny than the former, which usually emerge only postnatally (Hellbrügge 1960).

The technical breakthrough of push-pull cannula-

tion of the median eminence in freely-moving animals led to the demonstration of an ultradian pulsatility underlying the circadian pattern of hypothalamic neurohormone release—*e.g.* LHRH (Levine and Ramirez 1980), SRIF (Arancibia *et al.* 1984) and CRF (Ixart *et al.* 1987). There is recent evidence for a physiological role for ultradian pulsatility, for example, the circhoral (1 h) pulsatility in LHRH release, which is indispensable for the LH surge inducing ovulation (Knobil 1980).

As to the mechanisms of ultradian and circadian oscillators of neuroendocrine rhythms, neurobiological studies on selected animal models have shown that a phasic or bursting arrangement in single-unit electrical activity recorded in discrete neuronal populations is associated with a pulsatile release of their specific neurohormones, for example, LHRH-secreting neurons (Dufy *et al.* 1979) or oxytocin releasing neurons during suckling (Moos and Richard 1983).

On the other hand, circadian pacemakers appear to require more complex neuronal networks, and the prevailing concept is that they consist of a collection of interconnected and interacting subsystems (multioscillators) (Wever 1979). From a medical viewpoint it is interesting to note that none of the comparative studies conducted until now in different mammalian species, including primates, have shown species differences in these essentially subcortical mechanisms.

Role of the Suprachiasmatic Nucleus (SCN)

Among the various CNS-structures proposed as participants in the central machinery generating circadian behavioral and neuroendocrine rhythms, the one that most clearly satisfies the criteria for a circadian pacemaker is the SCN, which is a small heterogenous nucleus of some 10,000 neurons located in the hypothalamus, just above the optic chiasma. The salient experimental arguments favoring this concept include the following: (1) bilateral lesions of the SCN in the rat, mouse, hamster, ground-squirrel and rhesus monkey obliterated the circadian rhythm of general activity, feeding and drinking behaviours, body temperature, plasma ACTH, corticosterone or cortisol, TSH, prolactin and LH levels (for reviews see Krieger 1979, Assenmacher *et al.* 1987b); the lesion also disrupted the circadian release of melatonin by the mammalian pineal gland, which governs a number of neuroendocrine functions (Reiter 1986); (2) circadian fluctuations in electrical neuronal activity have been recorded both *in situ* in hypothalamic islands containing the SCN (Inouye and Kawamura 1979), and *in vitro* in hypothalamic sections including the SCN (Groos and Hendricks 1982); (3) circadian drinking behavior lost following SCN lesion was partially restored by intracerebral grafting of SCN-containing brain tissues (Drucker-Colin *et al.* 1984).

In spite of this impressive body of experimental data favoring the SCN as a major CNS controller of circadian rhythms, it may not be the sole source of circadian rhythmicity in the brain. A closer observation of rhythmic processes following SCN deletion has shown a persistence of circadian patterns *e.g.* motor activity, body temperature, plasma corticosterone, feeding, in laboratory animals such as rats, hamsters and squirrel-monkeys which were histologically confirmed to be SCN deprived (for review see Assenmacher 1987b). The removal of SCN was particularly ineffective when the circadian synchronization with the environment depended on synchronizers other than the light/dark cycle, for example the feeding schedule (Krieger *et al.* 1977, Boulos *et al.* 1980), which clearly points to the possible involvement of other CNS structures in circadian rhythmicity.

Role of CNS Structures Other than the Suprachiasmatic Nuclei

The pineal gland does not appear itself to be a circadian oscillator in mammals, but merely a transducer involved in circadian behavioral and neuroendocrine synchronization *via* its hormone melatonin. However, a few brain areas, sometimes neurochemically identified, have been shown (essentially in rat experiments) to be indispensable for the occurrence of circadian neuroendocrine rhythms.

These special CNS areas include the following (for review see Assenmacher *et al.* 1987b):

1) *The adrenergic/noradrenergic system*: Located in the pons (locus coeruleus) and in the medulla oblongata (dorsal A_2, and ventral A_1 nuclei), several populations of noradrenergic and adrenergic neurons project into the diencephalon and telencephalon via the noradrenergic ascending pathways; the hypothalamus receives its innervation essentially through the ventral noradrenergic bundle (VNAB) originating in the medullary A1 and A2 nuclei. Recent studies have shown that the suppression of hypothalamic catecholaminergic innervation by a discrete microinjection of the specific neurotoxin 6-hydroxydopamine into the VNAB leads to a disruption of the circadian ACTH rhythm, sparing short-lived ultradian fluctuations of reduced amplitude (Szafarczyk *et al.* 1985).

2) *The serotoninergic system*: Both the hypothalamic and limbic systems receive major serotoninergic innervations from the midbrain raphe nuclei. The prominent role of this system in controlling circadian neuroendocrine rhythms, including ACTH, TSH, and LH, was highlighted by the obliteration of these circadian rhythms by either stereotaxic lesions of the raphe nuclei, or pharmacological blockade of serotonin biosynthesis by the neurotoxin p-chlorophenylalanine (pCPA); the latter effect was reversed by a daily administration of 5-hydroxytryptophan, the serotonin precursor whose synthesis is blocked by pCPA (for review, see Assenmacher *et al.* 1987a and 1987b). Interestingly, the suppression of the specific serotoninergic innervation of the SCN by a local microinjection of the specific neurotoxin 5,7-dihydroxytryptamine, also led to a blockade of the circadian corticosterone rhythm (Williams *et al.* 1983, Banky *et al.* 1986).

3) *The GABA-ergic system*: GABAergic neurons are widespread in the CNS including the hypothalamus and the limbic system, and are found mainly as short interneurons. As in a variety of other brain functions, there is recent evidence that the GABAergic system provides inhibitory components that intervene in the central machinery governing circadian TSH (Jordan *et al.* 1983) and ACTH rhythms (for review see Assenmacher *et al.* 1987a, 1987b). For instance, a doubling of the hypothalamic GABA concentration by administration of a specific neurotoxicant (ethanol-amine-O-sulfate) directed against the catabolic enzyme of GABA, GABA-transaminase, blocked the ACTH rhythm at baseline levels, whereas infusion of the GABA antagonist picrotoxin blocked the rhythm at peak diurnal levels (Ixart *et al.* 1983a, Assenmacher *et al.* 1987a, 1987b).

4) *The limbic amygdala*: In view of the many an-

atomical and functional associations between the hypothalamus and the limbic system, the involvement of selected limbic areas in the central control of circadian neuroendocrine rhythms appears to be quite logical. In this line of research, a bilateral deletion of the basolateral amygdala induced a splitting of the circadian ACTH rhythm into short-lived ultradian fluctuations (for reviews see Assenmacher *et al.* 1987a).

Concluding Remarks

Since the adaptive role of most circadian rhythms is still unknown, there has been a tendency to consider them as curiosities rather than as basic processes of physiological regulation. More recently, the emergence of chronopathology, for example, the study of sleep disorders and digestive pathology associated with repeated environmental desynchronization of biological rhythms (for example, shift-work or transmeridian jet flights), and chronopharmacology (timing of drug administration) has led to the development of basic research on the mechanisms underlying circadian rhythms.

With reference to the mechanisms that generate circadian behavioural and neuroendocrine rhythms, that is the circadian organization of otherwise ultradian functional fluctuations, recent neurological studies have shown that whatever prominent role might be ascribed to the suprachiasmatic nuclei (SCN), a number of other brain areas, interacting directly or indirectly with the SCN, take part in a multioscillator system. The integrity of the system is essential for normal functioning of circadian neurorhythms, although a certain degree of vicariousness may exist when the system undergoes partial anatomical or functional deficiencies.

This circadian system involves complex neuronal networks interconnecting discrete regions of the brain stem, hypothalamus and limbic system. Consequently, as in the case of other highly integrated functions of the organism, there is no neurosurgical approach to the repair of circadian neurorhythm dysfunction yet envisaged, although neurosurgical procedures designed for other pathologies may disrupt circadian rhythms and their very strict mutual temporal relationships. Nevertheless, in the long run, it is probable that deeper insight into the central mechanisms of circadian neurorhythms will provide a valuable diagnostic insight for neurology and neurosurgery as well.

References

Arancibia S, Epelbaum J, Boyer R, Assenmacher I (1984) In vivo release of somatostatin from rat median eminence after K$^+$ infusion or delivery of nociceptive stress. Neurosci Lett 50: 97–112

Aschoff J (1981) Biological rhythms. Plenum, New York, 563 p

Assenmacher I, Szafarczyk A, Alonso G, Ixart G, Barbanel G (1987a) Physiology of neural pathways affecting CRH secretion. In: Ganong WF, Dallman MF, Roberts JL (eds) The hypothalamic-pituitary axis revisited. Proc New York Acad Sci 512: 149–161

— — Boissin J, Ixart G (1987b) CNS structures controlling circadian neuroendocrine and behavioral rhythms in mammals, Vol 3. Comparative physiology of environmental adaptations Pevet P (ed), Karger AG, Basel, 56–70

Banky Z, Halasz B, Nagy G (1986) Circadian corticosterone rhythm did not develop in rats 7 months after destruction with 5,7-dihydroxytryptamine of the serotoninergic nerve terminals in the suprachiasmatic nucleus at the age of 16 days. Brain Res 369: 119–124

Bargiello TA, Jackson FR, Young MW (1984) Restoration of circadian behavioral rhythms by gene transfer in Drosophila. Nature 312: 752–754

Boulos Z, Rosenwasser A, Terman M (1980) Feeding schedules and the circadian organization of behavior in the rat. Behav Brain Res 1: 39–65

Drucker-Colin R, Aguilar-Roblero R, Garcia-Hernandez F, Fernandez-Cancino F, Rattoni FB (1984) Fetal suprachiasmatic transplants. Diurnal rhythm recovery of lesioned rats. Brain Res 311: 353–357

Dufy B, Dufy-Barbe L, Vincent JD, Knobil E (1979) Etude électrophysiologique des neurones hypothalamiques et regulation gonadotrope chez le singe rhesus. J Physiol (Paris) 75: 105–108

Groos G, Hendricks J (1982) Circadian rhythm in electrical discharge of suprachiasmatic neurons recorded in vitro. Neurosci Lett 34: 283–288

Hellbrugge T (1960) The development of circadian rhythm in infants. Cold Spring Harbor Sympos Quantit Biol, New York, Long Island Biol Assoc, pp 311–323

Inouye ST, Kawamura H (1979) Persistence of circadian rhythmicity in a mammalian hypothalamic island containing the suprachiasmatic nucleus. Proc Nat Acad Sci USA 76: 5962–5966

Ixart G, Cryssogelou H, Szafarczyk A, Malaval F, Assenmacher I (1983a) Acute and delayed effects of picrotoxin on the adrenocorticotropic system of rats. Neurosci Lett 43: 235–240

— Szafarczyk A, Malaval F, Nouguier-Soule J, Assenmacher I (1983b) Persistence du couplage entre les rythmes circadiens des hormones hypophysaires chez des rattes évoluant en libre cours. C R Soc Biol, Paris 177: 58–64

— Barbanel G, Conte-Devolx B, Grino M, Oliver C, Assenmacher I (1987) Evidence for basal and stress-induced release of corticotropin releasing factor in the push-pull cannulated median eminence of conscious free-moving rats. Neurosci Let 74: 85–89

Jordan D, Poncet C, Veissiere M, Mornex R (1983) Role of GABA in the control of thyrotropin secretion in the rat. Brain Res 268: 105–110

Knobil E (1980) The neuroendocrine control of the menstrual cycle. In: Greer RO (ed) Recent progress in hormone research, Vol 36. Academic Press, New York, pp 53–88

Krieger DT (1979) Endocrine rhythms. Raven Press, New York, 372 p

— Hauser H, Krey L (1977) Suprachiasmatic nucleus lesions do not abolish food-shifted circadian adrenal and temperature rhythmicity. Science 197: 398–399

Levine JE, Ramirez VD (1980) In vivo release of luteinizing hormone-releasing hormone estimated with push-pull cannulae from the medio-basal hypothalami of ovariectomized steroid primed rats. Endocrinology 107: 1782–1790

Moore-Ede MC, Sulzman FM, Fuller CA (1982) The clocks that time us: Physiology of the circadian timing system. Harvard Univ Press, Cambridge

Moos F, Richard P (1983) Serotoninergic control of oxytocin release during suckling in the rat: opposite effects in conscious and anesthetized rats. Neuroendocrinology 36: 300–306

Reddy P, Zehring WA, Wheeler DA, Pirrotta V, Hadfield C, Hall JC, Rosbach M (1984) Molecular analysis of the period locus in drosophila melanogaster and identification of a transcript involved in biological rhythms. Cell 38: 701–710

Reiter R (1986) The pineal gland: an important link to the environment. News in Physiol Sci 1: 202–205

Szafarczyk A, Alonso G, Ixart G, Malaval F, Assenmacher I (1985) Diurnal-stimulated and stress-induced ACTH release in rats is mediated by central noradrenergic bundle. Am J Physiol E249: 219–226

Wever RA (1979) The circadian system of man. Springer, New York Heidelberg Berlin, 276 p

Williams JH, Miall-Allen VM, Klinowski M, Azmitia EC (1983) Effects of microinjections of 5-7-dihydroxytryptamine in the suprachiasmatic nuclei of the rat on serotonin reuptake and the circadian variations of corticosterone levels. Neuroendocrinology 36: 431–435

Correspondence: I. Assenmacher, Laboratory Endocrinological Neurobiology, UA 1197-CNRS, University of Montpellier II, F-34060 Montpellier, France.

Acta Neurochirurgica, Suppl. 47, 42–47 (1990)

The Vascular Supply of the Hypothalamus-Pituitary Axis

J. Lobo Antunes[1] and **K. Muraszko**[2]

[1] Department of Neurosurgery, University of Lisbon, Hospital de Santa Maria, Lisbon, Portugal
[2] Department of Neurosurgery, The Neurological Institute of New York, New York, U.S.A.

In 1930 Popa and Fielding described in the rat a specialized network of blood vessels, the hypothalamo-hypophyseal portal system, that connected the infundibulum (median eminence) and the sinusoids of the pars distalis of the pituitary gland. Its functional importance was not immediately understood. They found no venous drainage from the pituitary gland so they assumed that blood ran from the gland into the brain.

Later, Wislocki and King (1936) described the portal system in more detail. They found that the superior hypophyseal arteries gave rise to the capillary plexus of the median eminence. From this plexus originated the long portal vessels which ran along the pituitary stalk to terminate in a secondary plexus in the pars distalis. From here blood flowed through lateral hypophyseal veins into the systemic circulation. The major contribution of Green and Harris (1947) was to demonstate that the portal system constituted the anatomical framework through which the hypothalamus controlled the pituitary secretion.

Around the same time the concept of neurosecretion, that is the idea that nerve cells were able to secrete hormones and delivered them into the blood stream, was taking shape, thanks particularly to the morphological studies of the Scharrers (1940).

In Fig. 1 is outlined in a very simplified manner some of the neurovascular connections within the hypothalamic-pituitary axis.

Vascular Anatomy

The portal capillary system is present in all mammalian species that have been studied, although there are some variations in anatomical details. We will use as paradigm the vascular system of the rhesus monkey (Fig. 2)

following the landmark studies of Bergland and Page (1978). The arterial supply comes from branches of the internal carotid arteries. The median eminence receives its irrigation from the superior hypophyseal arteries, whereas the pars nervosa (or infundibular process) depends upon the inferior hypophyseal arteries. The middle hypophysial arteries (also called trabecular or loral arteries) go to the infundibular stem (pituitary stalk)

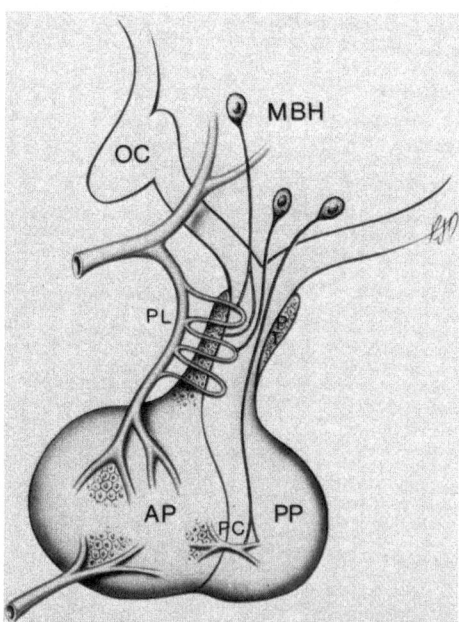

Fig. 1. Neurons in the medial basal hypothalamus (*MBH*) terminate in the pars nervosa or posterior pituitary (*PP*). Hormones or releasing factors liberated there may reach the adenohypophysis or anterior pituitary (*AP*) through short portal vessels (*PC*). Other neurons terminate in the inner plexus of the median eminence, and their products may reach the adenohypophysis through the long portal vessels (*PL*). Some nerve terminals may end in the cells of the pars tuberalis (*PT*). *OC* Optic chiasm. (From Antunes 1979)

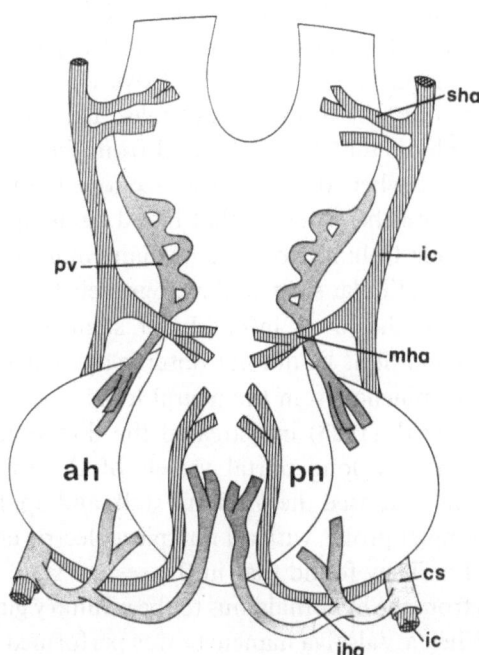

Fig. 2. Vascular supply of the pituitary gland. *ah* adenohypophysis; *pn* pars nervosa; *sha* superior hypophyseal artery; *ic* internal carotid; *mha* middle hypophyseal artery; *iha* inferior hypophyseal artery; *cs* cavernous sinus; *pv* portal vessel. Draining veins from both the anterior and posterior lobes join to drain into the cavernous sinus. (Adapted from Bergland and Page 1978)

and to the infundibular process (pars nervosa). In this species the adenohypophysis does not appear to have a direct arterial supply.

The superior hypophyseal arteries supply the capillary bed of the median eminence which is made of an outer shell, called the "external plexus", and an inner core of capillary coils, the "internal plexus". These project into the inner part of the infundibulum. It should be emphasized that the capillary networks of the various segments of the neurohypophysis—the infundibulum, the infundibular stem and the infundibular process—are inter-connected to form a common bed.

The capillary bed of the pars distalis is connected with the capillary bed of the infundibulum through the long portal vessels. The number of these is somewhat variable but there are always one or two running longitudinally on the anterior surface of the stalk. In addition the adeno- and neurohypophysis also communicate through short portal vessels.

The venous drainage is peculiar: the adenohypophysis does not have direct venous drainage into the cavernous sinus. Indeed its veins join similar veins from the neurohypophysis to form the confluent veins which drain into the cavernous sinus. In addition there are connections between the capillaries of the upper median eminence and the medial basal hypothalamus. It is likely that some of the blood of the neurohypophysis drains to the venous channels of the base of the brain.

Page (1982) unable to find a direct arterial supply to the adenohypophysis in other species such as mouse, rat, rabbit, cat, dog, and sheep. In man, however, there appears to be a dual arterial supply (Gorczyca and Hardy 1987), although the portal vessels provide the most important contribution. The lateral wings of the gland and the subcapsular peripheral zone may receive a contribution from penetrating vessels derived from the capsular arterial network. The "rete" is composed of branches of the inferior hypophyseal arteries, as well as vessels that come directly from the medial aspect of the intra-cavernous carotid—the capsular arteries of McConnell. Furthermore there may be vessels that arise directly from the inferior hypophyseal artery, as well as branches of the middle hypophyseal artery, which penetrate the anterior and superior surface of the gland.

In the dog there are no superior hypophyseal arteries but rather a system of arterial branches that form a plexus around the infundibulum (Dandy and Goetsch 1911).

Since section of the pituitary stalk has been performed as a therapeutic procedure in man, and as an experimental model in a number of animals, it is worth mentioning the vascular consequences of this procedure. The operation interrupts the long portal vessels and causes a variable degree of necrosis of the adenohypophysis estimated to be about 75% (Daniel and Prichard 1975). In addition, there may be areas of necrosis in the basal hypothalamus interpreted as venous infarctions, due to interruption of the venous drainage.

We studied this problem in the rhesus monkey and used a microsurgical transorbital approach (Antunes et al. 1980). In contrast to the results of others (Adams et al. 1963) we observed that in some animals there was no necrosis of the adenohypophysis and in the remaining, the average area of infarction was around 20%. This suggests not only that the short portal veins may play a more important role than was previously suspected, but also that the adenohypophysis may receive a direct arterial supply.

Direction of Blood Flow in the Hypothalamus-Pituitary Axis

One of the most fascinating questions pertaining to the vascular physiology of the hypothalamic-pituitary unit is the direction of the blood that flows through it. As mentioned before, Popa and Fielding (1930) believed that blood circulated from the hypophysis to the brain. This concept was first challenged by Wislocki and King (1936) and by Morato (1939) using injection techniques. The first demonstration "*in vivo*" that the direction of flow was indeed the reverse was provided by Houssay *et al.* (1935) and subsequently Green and Harris (1947) and Worthington (1955).

It is well documented that a short feedback loop operates between the hypothalamus and pituitary (see Antunes 1979). The demonstration of bidirectional flow within the portal system, or some form of circular flow between both structures will give additional support to that concept.

Page (1982) addressed this question with "*in vivo*" studies in the pig. He exposed the median eminence,

pituitary stalk, and pituitary gland, and found that blood ran from the median eminence to the pars distalis. In some instances the injected dye also passed from the neural lobe to the pars distalis. This area of adenohypophysis that received blood from the neural lobe increased when the stalk was sectioned. In the neurohypophysis he observed that blood circulated in an anterograde fashion from the median eminence to the upper infundibular stem, and retrogradely from the neural lobe to the lower infundibular stem. In normotensive conditions he did not observe dye reaching the median eminence from the neural lobe.

Antunes *et al.* (1983) investigated the direction of blood flow in the long portal vessels of the rhesus monkey. They exposed the pituitary stalk and applied a micro Doppler probe, with a 1 mm piezoelectric crystal around it. They found that in the resting state the blood ran from the hypothalamus to the pituitary gland (Fig. 3). When a Valsalva maneuver was performed the flow came to a halt, although actual reversal of flow could not be demonstrated. It is possible however that this technique is not sensitive enough to document transient small changes of flow.

The vascular architecture and the ample communications between the neurohypophysis and the adenohypophysis allow circulation of blood between these structures, and in certain circumstances, there may be reversal of the direction. In normal situations blood may run from the median eminence to the adenohypophysis through the long portal vessels and then reach the neural lobe through the short portal veins or the confluent veins. From there it could go up the infundibulum again, and down through the long portal vessels. Further support to this concept that the pituitary

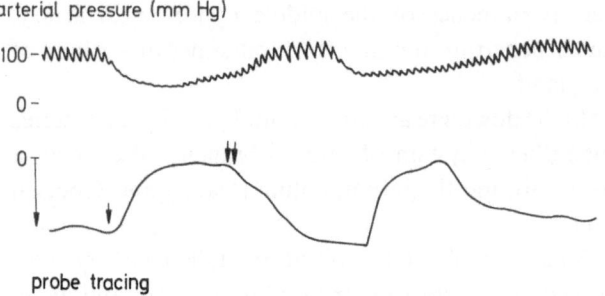

arterial pressure (mm Hg)

probe tracing

Fig. 3. Doppler recording of blood flow in long portal vessels, showing flow from the hypothalamus to the pituitary gland with a Valsalva maneuver (arrow): the recorded flow velocity dropped to zero, indicating no flow within the vessel. When the Valsalva maneuver was interrupted (double arrows) the flow returned to normal. The upper tracing displays the arterial pressure recording. (From Antunes *et al.* 1983)

Table 1. *Hypothalamus-pituitary blood flow*

Author	Species	Technique	Results (ml/100 g · min)		
			ME	AH	PH
Goldman	rat	^{86}Rb	—	59	360
David	rat	^{86}Rb	—	76	505
Porter	rat	H_2 clearance	—	37/170*	—
Lichardus	rat	^{125}I antipyrine	106	120	236
Page	sheep	microsphere	461	35	436
Muraszko	dog	microsphere	114/87**	48	625
Kemeny	rat	H_2 clearance	—	65/90*	—
Lees	rat	^{14}C I antipyrine	103	67/105*	133

ME Median eminence, AH adenohypophysis, PH posterior hypophysis.
 * Two components—"slow" and "fast".
** Two segments—ant. and post. m.e.

may secrete into the brain has come from the studies of Bergland and Page (1978) who measured the concentration of several pituitary hormones in various vascular channels which supply and drain blood from the hemispheres of the brain.

Blood Flow in the Hypothalamo-Pituitary Unit

The measurement of the blood flow to the hypothalamus and pituitary gland raises a number of methodological and technical questions, which are beyond the scope of this paper. They are related to the size of the gland, the heterogeneity of its composition, and its location. A number of techniques using fractional indicators, local hydrogen clearance, microspheres and autoradiography have been employed. The results of some of these studies are summarised in Table 1. Although there is some variation in the values obtained in different species, and within the same species, depending on the techniques used, some general statements can be made. In all studies, the posterior lobe has much higher values than the adenohypophysis.

In regard to the blood flow to the adenohypophysis Lees and Pickard (1987) recently called attention to the fact that its perfusion pressure (AHPP) was equal to the difference between the pressure in the portal veins (PVP) and within the sella (intrasellar pressure ISP), assuming of course, that the adenohypophysis does not have any direct arterial bloody supply. Since it is unlikely that the pressure in the portal vessels exceeds the peripheral venous pressure, it should be

Table 2. *Blood flow after stalk section (ml/100 g·min ± SEM)*

	Control	30 Cn*	15 SS**	30 SS***
Ant Pit	36 ± 4.5	26.7 ± 5.9	17.7 ± 4.9	11.0 ± 5.3
Post Pit	565.2 ± 60.1	346.9 ± 119.4	78.0 ± 10.3	70.0 ± 10.8
Ant Hyp	36.0 ± 2.0	35.2 ± 7.2	28.0 ± 4.9	30.7 ± 7.9
Post Hyp	38.5 ± 3.9	43.7 ± 9.3	39.8 ± 7.3	33.6 ± 6.9

* 30 minutes after craniotomy.
** 15 minutes after stalk section.
*** 30 minutes after stalk section.

lower than 10–15 mmHg. Thus, the AHPP is considerably lower than the cerebral perfusion pressure, and could easily be affected by any increase in the intrasellar pressure, as a result of an expanding lesion or rise in the pressure within the cavernous sinus.

When the microsphere technique was used the values obtained for the adenohypophysis were 35 ml/100 g min in the sheep (Page 1982) and 48 in the dog (Muraszko et al., unpublished observations). This is intriguing since in both instances spheres measuring approximately 15 μm in diameter were used, and they should have been trapped in the neurohypophyseal capillary bed, if one accepts the results of the morphological studies that seem to demonstrate the absence of direct arterial supply to the anterior lobe. One possible explanation is that smaller spheres may actually "escape"; another, is that there are indeed arteries going directly to the adenohypophysis which are not visualized by the techniques currently used to investigate the vascular anatomy.

We have also studied the flow to the gland in the dog following section of the pituitary stalk, using a microsurgical subtemporal exposure (Fig. 4). The results are briefly summarized in Table 2. Following the section there was a reduction in flow of both the anterior and the posterior lobes, particularly of the latter, which felt to 10% of the preoperative values. It is of interest to note that the flow to the anterior hypothalamus (which included the median eminence) and posterior hypothalamus were not altered.

Porter using the hydrogen clearance technique found that the blood flow in the adenohypophysis was not homogeneous: there was an area of fast flow in the centre of the gland, which probably received a contribution from the short portal vessels, and a slower flow in the periphery. Similar results were obtained recently by Lees et al. (1988).

It is worth emphasizing the high values found in all segments of the neurohypophyseal bed. In the sheep

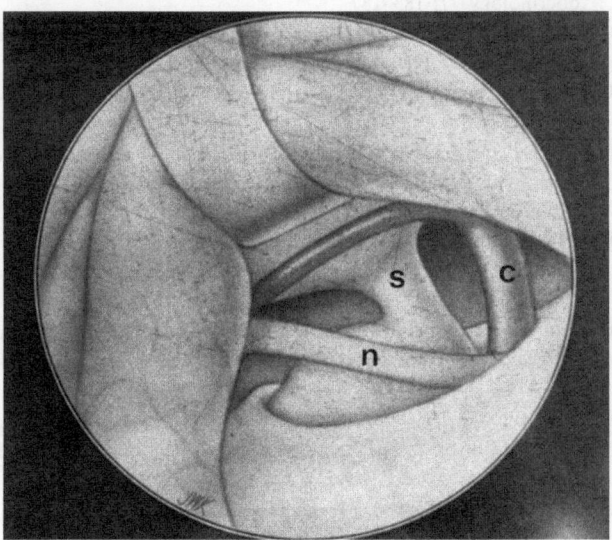

Fig. 4. Microsurgical subtemporal exposure of the pituitary stalk (*s*). The temporal lobe is elevated. *c* Carotid artery; *n* oculomotor nerve

Table 3. *Regional blood flow in the dog—ml/100 g · min (± SEM)*

Anesthetic agent	Ant. pit.	Post pit.	Ant. hypoth.	Post hypoth.	Cortical grey matter	Hemisph. white matter	Choroid plexus	Thyroid gland	Parotid gland
Ketamine	48.36 (± 11.93)	625.98 (± 73.42)	87.78 (± 7.09)	114.20 (± 11.96)	215.55 (± 29.05)	54.97 (± 6.19)	438.76 (± 48.08)	100.1 (± 16.1)	86.1 (± 17.8)
Pentobarbital	34.32 (± 7.38)	731.96 (± 139.39)	30.82 (± 3.21)	34.50 (± 4.71)	33.74 (± 3.37)	24.77 (± 2.51)	387.27 (± 54.11)	78.6 (± 11.2)	14.2 (± 1.4)

Page *et al.* (1981) encountered similar values for the median eminence and neural lobe, which further supports his concept of an anatomical and functional unity between these two structures. In the dog we divided the basal hypothalamus into two segments and found that the anterior one, which contained the median eminence, had a statistically lower value than the posterior one (Table 3). Among the neural structures we analysed, only the choroid plexus had flows comparable to the neural lobe.

As noted by Page (1982) the very high flow to the neurohypophysis reflect both the high metabolic demands of the neurosecretory processes and the ability to provide a rapid delivery of the hormonal substances.

The question of the reactivity of the vascular bed of the hypothalamo-pituitary axis was also addressed by Page *et al.* (1981). In their experiments when the $PaCO_2$ was lowered there was a decrease in flow of both the median eminence and neural lobe. When the $PaCO_2$ was increased, however, the median eminence blood flow did not change and the flow in the posterior lobe was only increased on average by 23% of the basal values, whereas the increase in other neural structures was much more pronounced. In addition, they did not find any significant alteration in flow when the blood pressure was raised or lowered beyond the limits of autoregulation.

In our own experiments (Table 3) we used two different anesthetic techniques. The values obtained for the anterior and posterior hypothalamus as well as for the cortical gray and hemispheric white matter were significantly higher with ketamine than with pentobarbital. There was however, no significant difference between the anterior and posterior pituitary lobes, which suggests that at least in the dog, the blood flow to the gland is under different regulatory mechanisms than the hypothalamus. Recently it has been shown in the rat that bromocriptine reduces the adenohypophyseal blood flow (Kemeny 1985b).

References

Adams JH, Daniel PM, Prichard MML (1963) Volume of the infarct in the anterior lobe of the monkey's pituitary gland shortly after stalk section. Nature 198: 1205–1206

Antunes JL (1979) Neural control of reproduction. Neurosurgery 5: 63–70

— Louis K, Cogen P, Zimmerman EA, Ferin M (1980) Section of the pituitary stalk in the rhesus monkey: II. Morphological studies. Neuroendocrinology 30: 76–82

— Murasko K, Stark R, Chen R (1983) Pituitary blood flow in primates: a Doppler study. Neurosurgery 12: 492–495

Bergland RM, Davis SL, Page RB (1977) Pituitary secretes to brain (experiment in sheep) Lancet 2: 276–278

— Page RB (1978) Can the pituitary secrete directly to the brain? (Affirmative anatomical evidence). Endocrinology 102: 1325–1338

Dandy WE, Goetsch E (1911) The blood supply of the pituitary body. Am J Anat 11: 137–150

Daniel PM, Prichard MML (1975) Studies of the hypothalamus and the pituitary gland. Acta Endocrinol 80 [suppl]

David MA, Csernay L, Laszlo FA, Kovacs K (1985) Hypophyseal blood flow in rats after destruction of the pituitary stalk. Endocrinology 77: 183–187

Goldman H, Sapirstein LA (1958) Determination of blood flow to the rat pituitary gland. Am J Physiol 194: 433–435

Gorczyca W, Hardy J (1987) Arterial supply of the human anterior pituitary gland. Neurosurgery 20: 369–377

Green JD, Harris GW (1947) The neurovascular link between the neurohypophysis and adenohypophysis. J Endocrinol 5: 136–146

Houssay BA, Biasotti A, Sammartino R (1935) Modifications fonctionnelles de l'hypophyse après lésions infundibulo-tubériennes chez le crapaud. C R Soc Biol (Paris) 120: 725–727

Kemeny AA, Jakubowski JA, Jefferson AA, Pasztor E (1985a) Blood flow and autoregulation in rat pituitary gland. J Neurosurg 63: 116–119

— — Pasztor E, Jefferson AA, Wojcikiewicz R (1985b) Reduction of blood flow in the adenohypophysis of rats by bromocriptine. J Neurosurg 63: 120–124

Lees PD, Pickard JD (1987) Hyperprolactinaemia, intrasellar pituitary tissue pressure, and the pituitary stalk compression syndrome. J Neurosurg 67: 192–196

— Richards HK, Pickard JD (1988) Changes in perfusion characteristics and blood flow in pituitary tumours. Adv Bio Sci 69: 457–460

Lichardus B, Albrecht I, Ponec J, Linhart L (1977) Water deprivation for 24 hours increases selectively blood flow in the posterior pituitary of conscious rats. Endocrinol Exp 11: 99–104

Morato MJX (1939) The blood supply of the hypophysis. Anat Rec 74: 297–320

Muraszko K, Chen RYZ, Antunes JL: To be published

Page RB (1982) Pituitary blood flow. Am J Physiol 243: E427–442

— (1982) In vivo recording of the direction of pituitary blood flow. Presented at the 64th Annual Meeting of the Endocrine Society

— Funsch DJ, Brennan RW, Hernandez MJ (1981) Regional neurohypophyseal blood flow and its control in adult sheep. Am J Physiol 241: R36–43

Popa G, Fielding V (1930) A portal circulation from the pituitary to the hypothalamic region. J Anat 65: 88–91

Porter JC, Dhariwal APS, McCann SM (1967) Quantitative evaluation of local blood flow of the adenohypophysis in rats. Endocrinology 80: 583–598

Scharrer E, Scharrer B (1940) Secretory cells within the hypothalamus. Res Publ Assoc Res Nerv Ment Dis 20: 170–174

Wislocki GB, King LS (1936) The permeability of the hypophysis and hypothalamus to vital dyes, with a study of the hypophyseal vascular supply. Am J Anat 58: 421–472

Worthington WC Jr (1955) Some observations on the hypophyseal portal system in the living mouse. Bull Johns Hopkins Hosp 97: 343–357

Correspondence: Prof. J. Lobo Antunes, Department of Neurosurgery, University of Lisbon, Hospital de Santa Maria, 1600 Lisbon, Portugal.

Acta Neurochirurgica, Suppl. 47, 48–57 (1990)

A Clinical Update on Hypothalamic-Pituitary Control

M. D. Page, J. Webster, C. Dieguez, R. Hall, and **M. F. Scanlon**

Neuroendocrine Unit, Department of Medicine, University of Wales College of Medicine, Heath Park, Cardiff, Wales

Introduction

Over the last three decades major advances in the understanding of hypothalamic-pituitary control mechanisms have led to important diagnostic and therapeutic developments. The early work of Geoffrey Harris, which established the concept of hypothalamic-pituitary control via the microvascular portal system was confirmed by the isolation by Guillemin and Schally in the late 1960's and early 1970's, of a variety of chemical mediators of these actions, the hypothalamic regulatory peptides. In more recent times the two most elusive hypothalamic regulatory peptides, corticotrophin releasing factor and growth hormone releasing factor (Table 1), have been isolated and characterised through the efforts of Vale and Guillemin. In this brief review we outline some of the more recent developments which have led to improvements in clinical management of the various endocrine conditions associated with disturbances of specific anterior pituitary hormones.

Table 1. *Major hypothalamic neuropeptides*

	Source	Amino acids	Actions
TRH	1968 Porcine, Ovine	3	↑ TSH, PRL (GH)
LHRH	1969 Porcine	10	↑ LH, FSH (GH)
SRIF	1970 Ovine	14, 28	↓ GH, TSH (PRL)
CRF	1981 Ovine	41	↑ ACTH (GH)
GRF	1982 Pancreatic tumour	40, 44	↑ GH (PRL)

Prolactin (PRL)

With present knowledge of the relatively benign course of most prolactinomas it is tempting to withhold therapy in asymptomatic patients, or those with minimal symptoms, particularly if fertility is not a problem. However, numerous studies of hyperprolactinaemic women have now demonstrated decreased density of cortical and trabecular bone using direct photon absorptiometry or CT scanning (Klibanski *et al.* 1980, Schlecte *et al.* 1983, Cann *et al.* 1984, Koppelmann *et al.* 1984, Klibanski and Greenspan 1986) and, in some, an increase in bone density when the condition is treated (Table 2). It is not yet fully established whether hyperprolactinaemia *per se* is involved in the decreased bone density or whether its effects are mediated by lowered plasma oestrogen levels. The latter seems more likely since, in a recent study, women with raised PRL but normal oestrogen levels and normal menstruation did not have decreased bone density (Ciccarelli *et al.* 1988). In the light of this evidence, there is a growing view at present that hyperprolactinaemic amenorrhoea should be treated even in the absence of other symptoms or indications, because of the risk of subsequent osteoporosis.

Medical therapy with bromocriptine remains the mainstay of treatment in most centres and alternative dopamine agonists, with fewer side effects, can be tried if there is bromocriptine intolerance. Such drugs include metergoline, dihydroergocryptine, lisuride and terguride. The longer acting and more potent agents pergolide and cabergoline hold particular promise, especially the latter drug which has a duration of action of several days following oral administration (Ferrari *et al.* 1986). The results of long term clinical trials of these two latter agents are awaited at present. A major

Table 2. *Bone density in hyperprolactinaemia*

No. patients	No. controls	Site	Method	Finding	Source
14	16	distal radius	direct photon absorptiometry	decreased	Klibanski *et al.* 1980
23	29	distal radius	direct photon absorptiometry	decreased	Schlechte *et al.* 1983
18	14 (untreated)	distal radius	direct photon absorptiometry	treatment with bromocriptine increased	Klibanski and Greenspan 1986
13	13	vertebral bodies	CT Scan	decreased	Koppelman *et al.* 1984
9	50	vertebral bodies	CT Scan	decreased	Cann *et al.* 1984

problem with dopamine agonist drugs is that treatment has to be over many years in the majority of patients if continuous cyclical behaviour and oestrogenisation are to be maintained. Although there is some evidence to indicate that long term dopamine agonist therapy contributes to the resolution of prolactinomas in a small proportion of patients, few groups have obtained as favourable results as Winkelmann and colleagues (1985). They treated 77 patients with prolactinomas, 10 with microadenomas and 67 with macroadenomas, with bromocriptine for 3–10 years and found persistent normoprolactinaemia after drug withdrawal in 5/10 patients with microadenomas and 7/67 macroadenomas. Overall, perhaps between three and ten per cent of patients go into complete remission each year during the course of dopamine agonist therapy.

An alternative treatment is partial hypophysectomy or microadenomectomy of prolactin-secreting tumours, and attitudes towards this approach have fluctuated in recent years. Firstly, the diagnosis of prolactinoma has to be established as accurately as possible. This can prove difficult since the normal gland can cause misleading appearances on high resolution CT scanning and very small adenomas can be missed completely. Dynamic testing of TSH and PRL release using TRH and domperidone is simple, safe and can be performed rapidly on an outpatient basis. In general, the results can be very helpful in supporting or excluding the presence of autonomous hyperprolactinaemia (Scanlon *et al.* 1985). The initial encouraging results of Hardy were tempered by his subsequent publication of around 50% recurrence rate at 5 years (Serri *et al.* 1983) which virtually precluded this approach as a meaning-

ful therapeutic manoeuvre. Fortunately other groups went on to publish their long term surgical series (Table 3) and whilst there is a definite recurrence rate, this is generally much lower than reported by Hardy's group. Our own experience in Cardiff in relation to surgery and recurrence is shown in Tables 4 and 5. At present a balance is being achieved between the use of dopamine agonist therapy and surgery. All options

Table 3. *Recurrence of micro-prolactinomas after surgery*

No. patients	Mean follow up (years)	Recurrence rate %	Source
24	6.2	50	Serri *et al.* 1983
55	1.7	5.5	Bertrand *et al.* 1983
50	3.1	16	Fahlbusch *et al.* 1984
50	3.1	16	Buchfelder *et al.* 1985
23	5.0	9	Thomson *et al.* 1985
15	2.7	0	Scanlon *et al.* 1985
20	5.0	40	Schlechte *et al.* 1986
16	4.0*	0	Bevan *et al.* 1987

* Median.

Table 4. *Cure rate in relation to surgeon's estimation of size*

Size (mm)	No. patients	Post-op PRL < 420 mU/l No. (%)
0–4	18	11 (61)
5–9	21	20 (95)
10–19	26	21 (81)
>20	7	4 (57)

Table 5. *Post-operative recurrence of prolactinomas. (Cardiff)*

Recurrence rate 4/72 (5%)

Duration of follow up 3–108 months (mean 39)

Post-op PRL (mU/l)	Time to recurrence (months)	Dynamic tests post-operatively
331	1	abnormal
299	1	abnormal
60	2	abnormal
64	30	abnormal

Cured but abnormal dynamic tests 7/72.

should be discussed fully with each patient and therapeutic decisions made on an individual basis.

The aetiology of prolactinomas remains largely unknown. In certain strains of rats and mice lactotroph hyperplasia and prolactinomas are readily induced by oestrogen administration. New arteries supplying the anterior lobe are prominent in this situation, the degree of arteriogenesis paralleling the sensitivity of the strain to oestrogen-induced tumour formation (Weiner *et al.* 1985). The arterial blood may dilute out portally-supplied dopamine and contribute to lactotroph hyperplasia and tumour formation. Certain strains of rats with spontaneous or oestrogen-induced PRL-secreting pituitary tumours show increased oncogene mRNA in their tumours. mRNA for the oncogenes c-fos, c-H-ras and c-myc (multiple species) have been described (White and Carlson 1987). Hence it is possible that increased expression of certain oncogene products, such as growth factors, their receptors or intracellular mediators might be involved in the evolution of prolactinomas in animals and in man.

In man, genetic factors can predispose to the development of prolactinomas as seen in the multiple endocrine neoplasia (MEN) Type I syndrome. Prolactinomas are common in certain local variants of this syndrome (Bear *et al.* 1985). No clear association of prolactinomas and oral contraceptive medication has emerged from several critical epidemiological studies. The cure of most prolactinomas by adenomectomy argues against any primary hypothalamic defect being involved in their pathogenesis. Teasdale (1983) was one of the first to suggest that prolactinomas might arise as a result of an anomalous arterial supply to the anterior lobe from the carotid arteries via the hypophyseal arteries rather than the usual hypothalamic portal plexus. This arterial supply could dilute the portal do-

pamine to a segment of the anterior lobe and lead to lactotroph hyperplasia and adenoma formation. Like Teasdale, we have observed several patients who, at operation for a prolactinoma, were found to have an abnormal artery supplying the anterior lobe, but no prolactinoma. Ligation of the artery was associated with a gradual fall in PRL levels over several weeks. Pre-operative neuroendocrine tests on these patients have shown atypical results with a normal pattern of TSH and PRL response to domperidone. It remains possible that patients with an anomalous arterial supply to the anterior lobe might be more sensitive to the effects of oestrogens in a combined process of tumourigenesis.

Growth Hormone (GH)

Our understanding of GH control has also increased considerably over the last decade (Dieguez *et al.* 1988) and various important clinical trials are presently under way based on several new therapeutic approaches. Long acting somatostatin analogues such as the Sandoz analogue SMS 201-995 have now replaced dopamine agonists as the first line of medical therapy in the treatment of patients with acromegaly. Several reports have demonstrated the efficacy of 8 or 12 hourly subcutaneous regimes of SMS administration, at dosage levels varying from 100 to 500 µg, in suppressing pathologically elevated GH levels into the normal range with parallel reduction in circulating IGF-1 levels (Fig. 1). The drug is well tolerated and its administration is associated with clinical improvement in the majority of treated patients (Page *et al.* 1987). There are limited data showing that the combined administration of somatostatin analogues and bromocriptine may lead to even further suppression of GH than when either agent is administered alone.

Surgical removal of a somatotroph adenoma re-

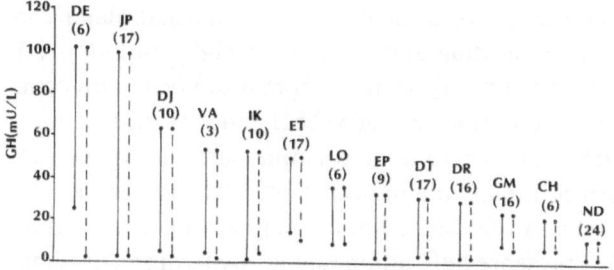

Fig. 1. GH suppression by SMS 201 995 in 13 acromegalics from mean pre-treatment values on Day 1 (solid line) and most recently (interrupted line). Duration of treatment in months shown in parentheses

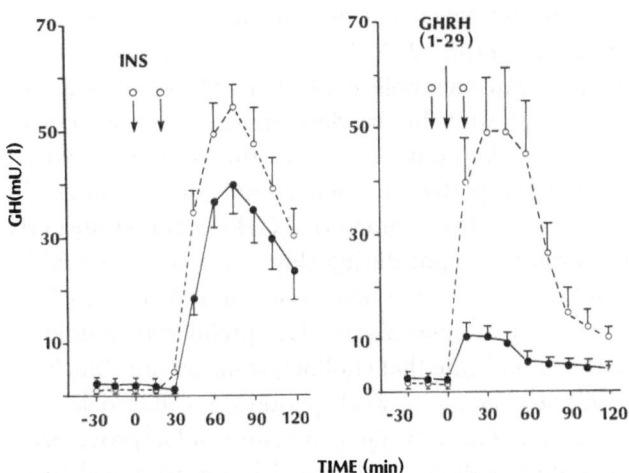

Fig. 2. Effect of atropine on GH response to GHRH and insulin-induced hypoglycaemia in normal subjects.
Atropine, ●; placebo, ○

mains the treatment of choice in acromegaly but when there are any contraindications or when surgery is not completely successful, the use of somatostatin is helpful. The effects of long term somatostatin administration on somatotroph adenoma size are unclear and the data are conflicting. Overall it appears that there may

be a slight reduction in size in a few patients with acromegaly, but the results are clearly very different to the dramatic shrinkage which can occur in many patients with large prolactinomas treated with dopamine agonists. It is reassuring however that there does not appear to be any further increase in size in acromegalic tumours in patients treated with somatostatin over the duration studied and we await the results of more long term prospective studies in relation to tumour size.

The other major area of physiological and clinical interest concerns the important role of cholinergic mechanisms in the control of GH synthesis and release. It has been known for many years that hypothalamic cholinergic pathways have been involved in GH control but the great sensitivity of the somatotroph to alterations in cholinergic mechanisms has only recently been appreciated. Cholinergic muscarinic receptor blockade with drugs such as atropine or pirenzepine usually completely abolishes the GH response to all known stimuli, including pharmacological doses of exgenous GHRH, with the exception of insulin-induced hypoglycaemia (Fig. 2).

In parallel with this, cholinergic activation with acetylcholinesterase blocking drugs such as pyridostigmine leads to enhancement of GH responsiveness to GHRH

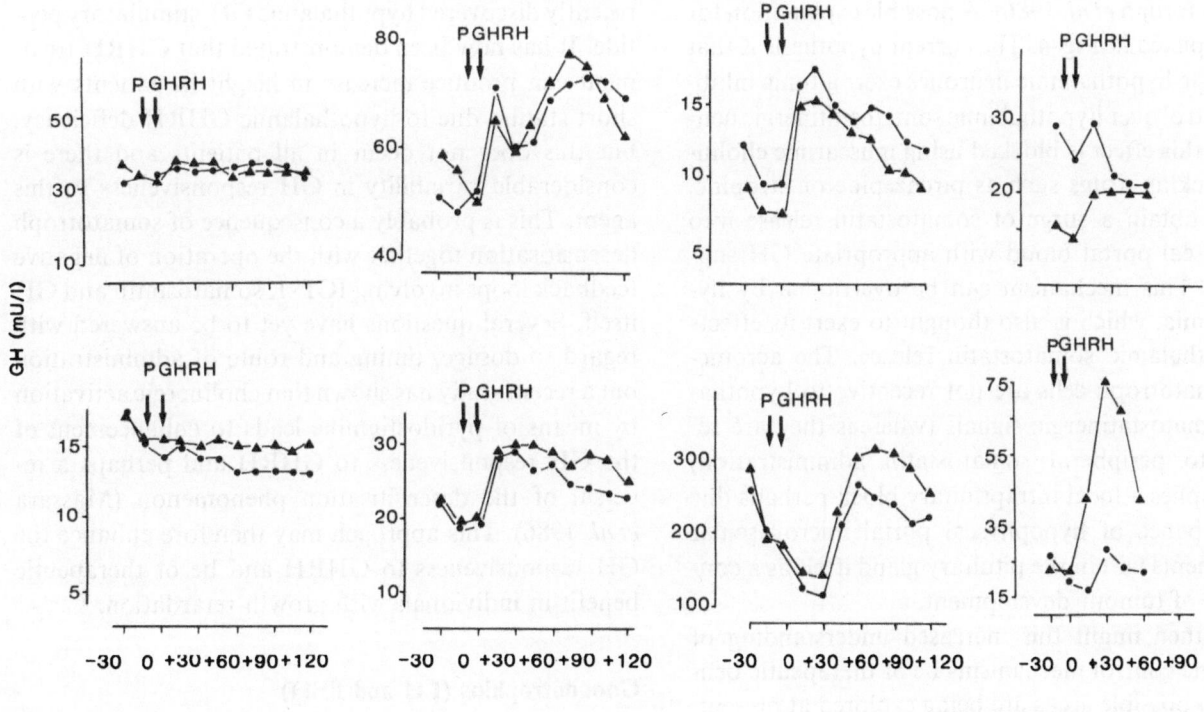

Fig. 3. GH response to GHRH and GHRH with pirenzepine (*P*) in eight acromegalic subjects (▲) GHRH alone: (●) with pirenzepine

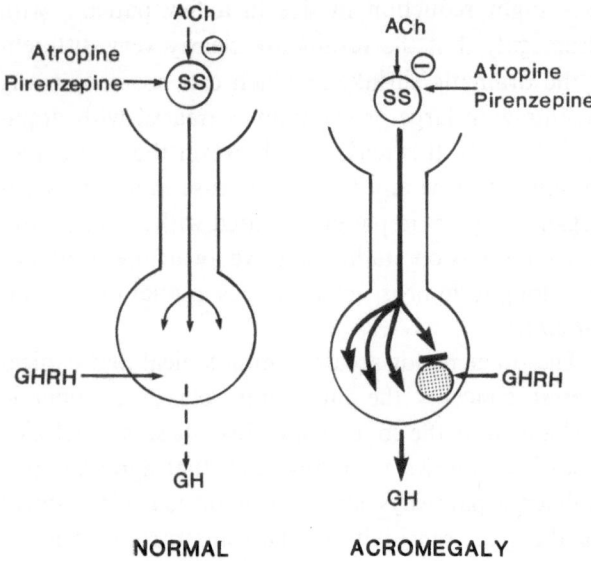

Fig. 4. The functional disconnection of adenomatous somatotrophs in acromegaly. A possible explanation for the different effects of cholinergic blockade on GH secretion in normals and diabetics compared with acromegalics

(Massara *et al.* 1986). Despite the potent action of muscarinic receptor blocking drugs to abolish GH release, these agents are totally ineffective in the medical management of acromegaly (Fig. 3). We have found that pirenzepine does not affect either basal, GHRH- or TRH-induced GH release in a large number of such patients (Jordan *et al.* 1986). A possible explanation for this is depicted in Fig. 4. The current hypothesis is that cholinergic hypothalamic neurones exert a tonic inhibitory control over hypothalamic somatostatinergic neurones. If this effect is blocked using muscarinic cholinergic blocking drugs such as pirenzepine or atropine, one will obtain a surge of somatostatin release into hypophyseal portal blood with appropriate GH suppression. This mechanism can be overridden by hypoglycaemia, which is also thought to exert its effects on hypothalamic somatostatin release. The acromegalic somatotroph cells are not receptive to hypothalamic somatostatinergic signals (whereas they are receptive to peripheral somatostatin administration) which implies a local intrapituitary block perhaps due to disturbance of hypophyseal portal microvascular arrangements within the pituitary gland itself as a consequence of tumour development.

How then might this increased understanding of cholinergic control mechanisms be of therapeutic benefit? Two possible areas are being explored at present. We have demonstrated that cholinergic muscarinic receptor blockade with pirenzepine can abolish the nocturnal GH surge in both normal subjects and in patients with insulin-dependent diabetes mellitus (Peters *et al.* 1986, Page *et al.* 1987). GH hypersecretion can lead to altered acute metabolic control in diabetes and has also been implicated in the development of a variety of microvascular complications in this condition, particularly retinopathy. In general, a safe and well tolerated means of abolishing nocturnal GH secretion (most GH is secreted at night during slow wave sleep) would be helpful in improving acute control and perhaps long term microvascular disease. Our preliminary studies in diabetics indicate that cholinergic muscarinic blockade at night over a one week period does indeed lead to reduction of the dawn phenomenon and improvements in acute metabolic control which are presumed to be consequent upon the quite marked GH inhibition which occurs (Atiea *et al.* 1989a, Atiea *et al.* 1989b). A possible benefit of this therapeutic approach, as opposed to the use of long acting somatostatin analogues, is the sparing of the counterregulatory GH response to hypoglycaemia which occurs with cholinergic muscarinic blockade but not with somatostatinergic inhibition of GH. This might avoid the brittleness of control which can occur when somatostatin is given in insulin-dependent diabetes mellitus.

The second area where cholinergic control mechanisms may be therapeutically important is in the treatment of children with short stature with GHRH, the recently discovered hypothalamic GH-stimulatory peptide. It has now been demonstrated that GHRH treatment can produce increase in height in patients with short stature due to hypothalamic GHRH deficiency, but this does not occur in all patients and there is considerable variability in GH responsiveness to this agent. This is probably a consequence of somatotroph desensitisation together with the operation of negative feedback loops involving IGF-1, somatostatin and GH itself. Several questions have yet to be answered with regard to dosage, timing and route of administration but a recent study has shown that cholinergic activation by means of pyridostigmine leads to enhancement of the GH responsiveness to GHRH and perhaps a reversal of the desensitisation phenomenon (Massara *et al.* 1986). This approach may therefore enhance the GH responsiveness to GHRH and be of therapeutic benefit in individuals with growth retardation.

Gonadotrophins (LH and FSH)

Increases in our understanding of the physiology of gonadotrophin control over the last ten years have led to a variety of therapeutic advances. Though it was appreciated for many years that the hypothalamic peptide gonadotrophin-releasing hormone (LHRH) was a physiological regulator of gonadotrophin synthesis and release, it was not until the elegant studies of Knobil (1980) in primates that full significance of the underlying mechanisms involved was demonstrated. In monkeys with ablative hypothalamic lesions leading to hypogonadotrophic hypogonadism, a normal pattern of cyclical hormonal behaviour could be restored by pulsatile administration of LHRH peripherally. From this series of experiments it became clear that pulsatile endogenous LHRH release from the hypothalamus was absolutely critical to the normal rhythms of gonadotrophin secretion. This is because the gonadotroph cell is rapidly desensitised when exposed to continuous LHRH, leading to a fall in gonadotrophin output. The importance of this approach was established in humans when it was demonstrated that normal cyclical activity and menstruation could be restored to women with hypothalamic hypogonadism, amenorrhoea and infertility following administration of exogenous pulsatile LHRH (Mason *et al.* 1984). This approach uses subcutaneous administration of pulses of LHRH (usually about 15 µg every 90 minutes via a portable mini-infusion pump) and is now standard treatment for females with true hypogonadotrophic hypogonadism and infertility due to presumed LHRH deficiency. Use of this natural mechanism avoids the hyperstimulation syndrome and multiple pregnancies associated with human menopausal gonadotrophin and human chorionic gonadotrophin.

Fuller appreciation of the extent of gonadotroph desensitisation following chronic exposure to high concentrations of LHRH also led to unexpected therapeutic developments. Chronic infusion of LHRH or administration of super-agonist analogues of LHRH leads to rapid and reversible desensitisation of the gonadotroph (following an initial abrupt stimulation) causing reversible clinical hypogonadism in both males and females. Intra-nasal administration of super-agonist analogues of LHRH is now of therapeutic benefit in such diverse conditions as precocious puberty, fibroids, menorrhagia, endometriosis and hormone-dependent carcinomas of the prostate and breast. It may also have a role in contraception and as therapy for the premenstrual syndrome.

Adrencorticotrophic Hormone (ACTH)

Further understanding of hypothalamic control of ACTH and proopiomelanocortin (POMC) gene ex-

pression followed the isolation by Vale and colleagues (1981) of corticotrophin-releasing factor (CRF), a 41 amino acid peptide. It is now also established that there is very close interaction between both CRF and vasopressin in the control of POMC gene expression and corticotroph function. A distinct pathway has now been described for vasopressin producing cells arising in the parvocellular division of the paraventricular nucleus and projecting to the median eminence. This is an addition to the well established vasopressin supply to the posterior pituitary gland from the supraoptic and suprachiasmatic nuclei. The parvocellular paraventricular cell bodies contain specific vasopressin and CRF mRNA, the levels of which alter appropriately in response to glucocorticoid treatment or hypoadrenalism (Sawchenko *et al.* 1984, Wolfson *et al.* 1985, Robinson 1986, Gillies and Lowry 1986).

There have been numerous clinical studies of the application of CRF to the investigation of hypercortisolism. Predictably, patients with pituitary-dependent Cushing's disease show exaggerated ACTH and cortisol responses to CRF compared with normals, depressed patients, patients with primary adrenal disease and patients with the ectopic ACTH syndrome (Chrousos *et al.* 1985, Muller and Von Werder 1988). Whilst these differences are clear on a group mean basis, as with all tests of adrenal function, there can be considerable overlap between different aetiological groups on an individual basis, for example, occasional patients with ectopic ACTH syndromes show positive responses to CRF (Muller and Von Werder 1988). However, recent reports have suggested a more valuable indication for the use of CRF in the investigation of ACTH-dependent hypercortisolism. That is in association with selective bilateral inferior petrosal sinus sampling for measurement of ACTH levels (Landolt *et al.* 1986, Schulte *et al.* 1988). This investigative procedure can be used to determine the pituitary origin of secreted ACTH and also to indicate which side of the pituitary gland is the source of autonomous ACTH secretion. This information is valuable in relation to subsequent surgical exploration of the fossa. An interesting recent finding is the demonstration, using this approach, of PRL responses to CRF on the same side of the pituitary as the ACTH response in patients with Cushing's disease (Schulte *et al.* 1988). This suggests that many ACTH-producing pituitary tumours are also PRL-producing which is in accord with other recent evidence (Sherry *et al.* 1982, Yamaji *et al.* 1984, 1985, Schulte *et al.* 1988).

Thyroid Stimulating Hormone (TSH)

The TRH test has been used most commonly in the investigation of primary hyperthyroidism in that the presence of any significant TSH response to TRH virtually excludes this diagnosis. However this diagnostic approach has now been superseded by the development of ultra-sensitive assays for human TSH. Such reagent excess, two-site assays using either fluorescent, radioiodine or chemiluminescent labels have improved sensitivity dramatically and wide discrimination can now be achieved between hyperthyroid and euthyroid groups (Weeks *et al.* 1984). Quite clearly such TSH assays are also useful in the diagnosis of hyperthyroidism due to inappropriate TSH secretion when TSH levels may be in the normal range. Future studies may reveal that this syndrome is commoner than previously suspected. The use of TRH testing is still of value however in the investigation of hypothalamic-pituitary disease states in that an absent TSH response to TRH (without hyperthyroidism) is highly suggestive of an intrapituitary lesion whereas a delayed TSH response to TRH is more in keeping with either hypothalamic or pituitary stalk damage and is seen commonly in stalk-compression hyperprolactinaemia (Scanlon *et al.* 1986). Application of this new assay technology to other anterior pituitary hormones such as ACTH and gonadotrophins may also improve the clinical diagnostic armamentarium.

Growth Control in the Anterior Pituitary

The advances in understanding of the functional control of the anterior pituitary gland over the last decade have not been paralleled by similar advances in relation to pituitary growth control. Despite the fact that it has been known since the mid 1970's that dopamine agonism with drugs such as bromocriptine can lead to dramatic shrinkage of large prolactin-secreting tumours, there have been relatively few studies of the mechanism of action of such agents in relation to pituitary growth. Balanced growth of the anterior pituitary gland depends upon the coordinated production of specific anterior pituitary cell type growth factors, angiogenic factors and connective tissue growth factors. In particular, coordinated angiogenesis is crucial to the development and maintenance of normal hypophyseal portal microvascular relationships within the anterior pituitary gland. These relationships are probably disrupted when autonomous cellular growth occurs leading to adenoma formation. Compensatory growth of the anterior pi-

tuitary depends upon the interaction between the inhibitory target gland hormones, stimulatory hypothalamic regulatory hormones and a variety of growth factors acting in a paracrine or autocrine fashion. Growth factors such as IGF-1 should no longer be regarded simply as "circulating mediators" since their production has been demonstrated in a variety of tissues. It seems likely that the local production, action and metabolism of such factors within the tissue are more important determinants of cell growth. Certainly recent data have demonstrated that the anterior pituitary gland produces a variety of growth factors which may be involved in the regulation of cell growth and function (Table 6). However, their control and physiological roles are unknown as are their specific target cells. IGF-1 inhibits GH secretion and GH mNRA levels both basally and after stimulation with GHRH. Furthermore, IGF-1 receptors have been characterised on both anterior pituitary membranes and somatotroph adenoma cells. The IGF-1 gene is expressed in the anterior pituitary as indicated by *in situ* hybridisation of specific mRNA with IGF-1 cDNA probes (Melmed 1988).

EGF, at physiological concentrations, stimulates GH secretion from superfused normal rat anterior pituitary fragments without affecting TSH or PRL secretion (Ikeda *et al.* 1984). Also, EGF receptors are present on normal rat and human pituitary membranes but not on tumours taken from acromegalic patients (Michard *et al.* 1986). In contrast, in a tumour cell line (GH$_4$C$_1$ cells), EGF increases PRL gene transcription (Murdoch *et al.* 1982), stimulates PRL synthesis and inhibits GH synthesis (Kudlow and Kobrin 1984). Using another tumour cell line (GH$_3$D$_6$), Johnson *et al.* (1980) found that EGF alone did not stimulate cell division but inhibited cell growth stimulated by thyroid hormone-induced growth factor. However, their most striking finding was a marked alteration in the morphology of the cells after EGF treatment from a

Table 6. *Growth factors identified in the anterior pituitary*

- Insulin-like growth factors I and II
- Epidermal growth factor
- Chondrocyte growth factor
- Fibroblast growth factor
- Ovarian growth factor
- Glial growth factor
- Plasminogen activators
- Autostimulatory growth factor oestrogen insensitive
- Thyroid hormone-induced growth factor
- Mammary tumour cell growth factor

rounded to an elongated form. It has been suggested that EGF-like mitogens act within the pituitary to mediate continual cell turnover and permit population shifts in response to changes in the peripheral hormonal environment (Kudlow and Kobrin 1984).

PDGF inhibits PRL secretion in GH_3 cells and enhances GH release in GC cells. PDGF also affects cellular morphology leading to larger, flattened cells with angular borders which are more tightly adherent to the culture dish (Sullivan and Tashjian 1983). FGF increases the sensitivity of both thyrotrophs and lactotrophs to TRH (Baird *et al.* 1985) and inhibits, in a similar fashion to TRH and EGF, cell proliferation in GH_4C_1 cell lines (Schonbrunn *et al.* 1980). It is now clear that GHRH may also play an important part in the control of somatotroph cell growth as well as in the control of GH secretion. Using autoradiographic detection of ^{3}H-thymidine uptake and immunocytochemistry, it was shown that nanomolar concentrations of GHRH increased by up to 60% the total number of somatotrophs in vitro. Somatostatin caused partial inhibition of the stimulatory effect of GHRH (Billestrup *et al.* 1986). It has been reported recently that GHRH produced a very rapid activation of the c-fos oncogene in anterior pituitary cells, although this may be more related to cell differentiation than division (Billestrup *et al.* 1987). Of further interest is the recent report that the c-myc oncogene, which initiates the nuclear events of cell division, is expressed in cultured human adenomatous GH-secreting cells but not in other pituitary cell types studied (Isaac *et al.* 1987).

In summary, there is at present an accumulation of data which indicate the importance of stimulatory and inhibitory growth factors in the paracrine, and perhaps autocrine, regulation of anterior pituitary cell growth and function. There may well be several more specific pituitary growth factors which will be identified in the near future. Preliminary data are also emerging concerning the pattern of oncogene expression in human pituitary tumours. It is now necessary to relate growth factor production, oncogene expression and hypothalamic regulation to specific anterior pituitary cell populations and human pituitary adenomas. In the midst of these studies it should not be forgotten that coordinated vascular and supporting tissue growth is also crucial to the normal development and maintenance of anterior pituitary structure and function.

References

Atiea JA, Creagh F, Page MD, Owens DR, Scanlon MF, Peters JR (1989) Early morning hyperglycaemia in IDD. Acute effects of cholinergic blockade. Diabetes (in press)

— — — — — — (1989) Early morning hyperglycaemia in IDD: acute and sustained effects of cholinergic blockade. J Clin Endocrinol Metab (in press)

Baird A, Mormede P, Ying SY (1985) A nonmitogenic pituitary function of fibroblast growth factor: regulation of thyrotrophin and prolactin secretion. Proc Natl Acad Sci USA 82 (16): 5545-5549

Bear JC, Briones-Urbina R, Fahey JF, Farid NR (1985) Variant multiple endocrine neoplasia I (MEN I[Burin]): Further studies on non-linkage to HLAP₁. Hum Hered 35: 15–20

Bertrand G, Tolis G, Montes J (1983) Immediate and long-term results of transsphenoidal microsurgical resection of prolactinomas in 92 patients. In: Tolis G, Stefanis C, Mountokalakis T, Labrie F (eds) Prolactin and prolactinomas. Raven Press, New York, pp 441

Bevan JS, Adams CBT, Burke CW, Morton KE, Molyneux AJ, Moore RA, Esiri MM (1987) Factors in the outcome of transsphenoidal surgery for prolactinoma and non-functioning pituitary tumour, including pre-operative bromocriptine therapy. Clin Endocrinol 26: 541–556

Billestrup N, Mitchell RM, Vale W, Verma JM (1987) Growth hormone releasing factor induces c-fos expression in cultured primary pituitary cells. Mol Endocrinol 1 (4): 300–305

— Swanson LW, Vale W (1986) Growth hormone releasing factor stimulates proliferation of somatotrophs in vitro. Proc Natl Acad Sci USA 83 (18): 6854–6857

Buchfelder M, Lierheimer A, Schrell U (1985) Recurrence of hyperprolactinaemia detected in long-term follow-up of surgically normalized microprolactinomas. In: Auer LM, Leb G, Tscherne G, Urdle W, Walter GF (eds) Prolactinomas. Walter de Gruyter, Berlin New York, pp 183–187

Cann CE, Martin MC, Harry K, Genant HK, Jaffe RB (1984) Decreased spinal mineral content in amenorrheic women. JAMA 251: 626–629

Chrousos GP, Schürmeyer TH, Doppman J, Oldfield EH, Schulte HM, Gold PW, Loriaux DL (1985) Clinical applications of corticotropin releasing factor. Ann Intern Med 102: 344–358

Ciccarelli E, Savino L, Carlevatto V, Bertagna A, Isaia GC, Camanni F (1988) Vertebral bone density in non-amenorrhoeic hyperprolactinaemic women. Clin Endocrinol 28: 1–6

Dieguez C, Page MD, Scanlon MF (1988) Growth hormone neuroregulation and its alterations in disease states. Clin Endocrinol 28: 109–143

Fahlbusch R, Buchfelder M, Werder K (1984) Present status of surgical treatment of prolactinomas (and long term follow-up). In: Lamberts SW, Tilders FJH, Van der Veen EA, Assies J (eds) Trends in diagnosis and treatment of pituitary adenomas. Free University Press, Amsterdam, pp 121–132

Ferrari C, Barbieri C, Caldara R, Mucci M, Codecasa F, Paracchi A, Romano C, Boghen M, Dubini A (1986) Long-lasting prolactin-lowering effect of cabergoline, a new dopamine agonist, in hyperprolactinaemic patients. J Clin Endocrinol Metab 63: 941–945

Gillies GE, Lowry PJ (1986) Adrenal function. In: Lightman SL, Everitt BJ (eds) Neuroendocrinology. Blackwell Scientific Publications, Oxford, pp 360–388

Ikeda H, Mitsuhashi T, Kubota K, Kuzuya N, Uchimura H (1984) Epidermal growth factor stimulates growth hormone secretion from superfused rat adenohypophyseal fragments. Endocrinology 115: 556–558

Isaacs RE, Findell PR, Gertz BJ, Baxter JD (1987) C-myc gene expression in human growth hormone-secreting pituitary adenomas: regulation by glucocorticoids. Clin Res 35: 396A

Johnson LK, Baxter JD, Vlodausky I, Gospodarowicz D (1980) Epidermal growth factor and expression of specific genes: effects on cultured rat pituitary cells are dissociable from the mitogenic response. Proc Natl Acad Sci USA 77 (1) 394–398

Jordan V, Dieguez C, Valcavi R, Artioli C, Portioli I, Rodriguez-Arnao MD, Gomez-Pan A, Hall R, Scanlon F (1986) Lack of effect of muscarinic cholinergic blockade on the GH responses to GRF 1–29 and TRH in acromegalic subjects. Clin Endocrinol 24: 415–420

Klibanski A, Greenspan SL (1986) Increase in bone mass after treatment of hyperprolactinaemic amenorrhoea. N Eng J Med 315: 542–546

— Robert MD, Neer M, Beitins IZ, Chester Ridgway E, Zervas NT, McArthur JW (1980) Decreased bone density in hyperprolactinaemic women. N Engl J Med 303: 1511–1513

Knobil E (1980) The neuroendocrine control of the menstrual cycle. Recent Prog Hor Res 36: 53–58

Koppelman MCS, Kurtz DW, Morrish KA, Bou E, Susser JK, Shapiro JR, Loriaux DL (1984) Vertebral body bone mineral content in hyperprolactinaemic women. J Clin Endocrinol Metab 59: 1050–1053

Kudlow JE, Kobrin MS (1984) Secretion of epidermal growth factor-like mitogens by cultured cells from bovine anterior pituitary glands. Endocrinology 115: 911–917

Landolt AM, Valavanis A, Girard J, Eberle AN (1986) Corticotropin-releasing factor test used with bilateral, simultaneous inferior petrosal sinus blood-sampling for the diagnosis of pituitary-dependent Cushing's disease. Clin Endocrinol 25: 687–696

Mason P, Adams J, Morris D, Tucker M, Price J, Voulgaris Z, Van der Spuy ZM, Sutherland I, Chambers GR, White S *et al* (1984) Induction of ovulation with pulsatile luteinising hormone-releasing hormone. Br Med J 288: 181–185

Massara F, Ghigo E, Molinatti P, Mazza E, Locatelli V, Muller EE, Camanni F (1986) Potentiation of cholinergic tone by pyridostigmine bromide reinstates and potentiates the growth hormone responsiveness to intermittent administration of growth hormone releasing factor in man. Acta Endocrinol (Copenh) 113: 12–16

Melmed S (1988) Pituitary growth factors. In: Scanlon MF, Wass JAH (eds) Neuroendocrine perspectives. Springer, New York (in press)

Michard M, Birman P, Peillon F, Bression D (1986) EGF receptors present in normal rat and human pituitaries are absent in human PRL, GH, and non-secreting pituitary adenomas. First Inter Con Neuroendocrin, San Francisco, Abstract 146

Muller OA, von Werder K (1988) Diagnostic dilemmas in hypercortisolism: investigation and management. In: Scanlon MF, Wass JAH (eds) Neuroendocrine perspectives. Springer, New York (in press)

Murdoch GH, Potter E, Nicolaisen AK, Evans RM, Rosenfeld MG (1982) Epidermal growth factor rapidly stimulates prolactin gene transcription. Nature 300: 192–194

Page MD, Koppeschaar HPF, Dieguez C, Gibbs JT, Hall R, Peters JF, Scanlon MF (1987) Cholinergic muscarinic receptor blockade with pirenzepine abolishes slow wave sleep-related GH release in young patients with insulin dependent diabetes mellitus. Clin Endocrinol 26: 355–359

— Millward ME, Hourihan M, Hall R, Scanlon MF (1989) Long term treatment of acromegaly with a long-acting analogue of somatostatin, Ocreotide. Q J Med (submitted)

Peters JR, Evans PJ, Page MD, Hall R, Gibbs JT, Dieguez C, Scanlon MF (1986) Cholinergic muscarinic receptor blockade with pirenzepine abolishes slow wave sleep-related growth hormone release in normal adult males. Clin Endocrinol 25: 213–217

Robinson ICAF (1986) The magnocellular and parvocellular OT and AVP systems. In: Lightman SL, Everitt BJ (eds) Neuroendocrinology. Blackwell Scientific Publications, Oxford, pp 154–176

Sawchenko PE, Swanson LW, Vale WW (1984) Co-expression of corticotropin-releasing factor and vasopressin immunoreactivity in parvocellular neurosecretory neurons of the adrenalectomised rat. Proc Natl Acad Sci USA 81: 1883–1887

Scanlon MF, Peters JR, Thomas JP, Richards SH, Morton WH, Howell S, Williams ED, Hourihan M, Hall R (1985) Management of selected patients with hyperprolactinaemia by partial hypophysectomy. Br Med J 291: 1547–1550

— Rodriguez-Arnao MD, McGregor AM, Weightman D, Lewis M, Cook DB, Gomez-Pan A, Hall R (1981) Altered dopaminergic regulation of thyrotrophin release in patients with prolactinomas: comparison with other tests of hypothalamic-pituitary function. Clin Endocrinol 14: 133–143

Schlechte JA, Sherman BM, Chapler FK, Van Gilder J (1986) Long term follow-up of women with surgically treated prolactin-secreting pituitary tumours. J Clin Endocrinol Metab 62: 1296–1301

— — Martin R (1983) Bone density in amenorrhoeic women with and without hyperprolactinaemia. J Clin Endocrinol Metab 56: 1120–1123

Schonbrunn A, Krasnoff H, Westendorf JM, Tashjian AH (1980) Epidermal growth factors and thyrotrophin-releasing hormone act similarly on a clonal pituitary cell strain. Modulation of hormone production and inhibition of cell proliferation. J Cell Biol 85 (3): 786–797

Schulte HM, Allolio B, Günther RW, Benker G, Winkelmann W, Ohnhaus EE, Reinwein D (1988) Selective bilateral and simultaneous catheterization of the inferior petrosal sinus: CRF stimulates prolactin secretion from ACTH-producing microadenomas in Cushing's disease. Clin Endocrinol 28: 289–295

Serri O, Rasio E, Beauregard H, Hardy J, Somma M (1983) Recurrence of hyperprolactinaemia after selective transsphenoidal adenomectomy in women with prolactinoma. N Engl J Med 309: 280–283

Sherry SH, Guay AT, Lee AK, Hedley-White ET, Federmann M, Freidberg SR, Woolf PD (1982) Concurrent production of adrenocorticotropin and prolactin from two distinct cell lines in a single pituitary adenoma: a detailed immunohistochemical analysis. J Clin Endocrinol Metab 55: 947–955

Sullivan NJ, Tashjian AH (1983) Platelet-derived growth factor selectively decreases prolactin production in pituitary cells in culture. Endocrinology 113: 639–645

Teasdale G (1983) Surgical management of pituitary adenoma. In: Scanlon MF (ed) Clinics in endocrinology and metabolism. WB Saunders Co Ltd, London , pp 789–823

Thomson JA, Teasdale GM, Gordon D, McCruden DC, Davies DL (1985) Treatment of presumed prolactinomas by transsphenoidal operation: early and late results. Br Med J 291: 1550–1553

Vale W, Spiess J, Rivier C, Rivier J (1981) Characterisation of a 41-residue ovine hypothalamic peptide that stimulates secretion of corticotropin and B-endorphin. Science 213: 1394–1397

Weeks I, Sturgess M, Siddle S, Jones MK, Woodhead JS (1984) A high sensitivity immunochemiluminometric assay for human thyrotrophin. Clin Endocrinol 20: 489–495

Weiner RI, Elias KA, Monnet F (1985) The role of vascular changes in the aetiology of prolactin secreting anterior pituitary tumours. In: MacLeod RM, Thorner MO, Scapagnini U (eds) Prolactin basic and clinical correlates. Springer, New York, pp 641–653

White JD, Carlson HE (1987) Increased oncogene expression in rat pituitary tumours. In: Habener JF, Means AR, Ringold GM, Rosenfeld MG (eds) Programme and Abstracts: The Endocrine Society 69th Annual Meeting Indiana. Bethesda, Endocrine Society, Abst 650, pp 183

Winkelmann W, Allolio B, Deuss U, Heesen D, Kaulen D (1985) Persisting normoprolactinaemia after withdrawal of bromocriptine long-term therapy in patients with prolactinomas. In:

MacLeod RM, Thorner MO, Scapagnini U (eds). Prolactin: basic and clinical correlates. Liviana Press, Padova, pp 817–822

Wolfson B, Manning RW, Davis LG, Arentzen R, Baldina F (1985) Co-localisation of corticotrophin-releasing factor and vasopressin mRNA in neurons after adrenalectomy. Nature 315: 59–61

Yamaji T, Ishibaski M, Teramoto A, Fukushima T (1984) Hyperprolactinaemia in Cushing's disease and Nelson's syndrome. J Clin Endocrinol Metab 58: 790–795

— — — — (1985) Prolactin secretion by mixed ACTH-prolactin pituitary adenoma cells in culture. Acta Endocrinol (Copenh) 108: 456–463

Correspondence: M. D. Page, M.D., Neuroendocrine Unit, Department of Medicine, University of Wales College of Medicine, Heath Park, Cardiff, Wales.

Acta Neurochirurgica, Suppl. 47, 58–60 (1990)

Clinical Syndromes of the Hypothalamus

R. Fahlbusch, U. M. H. Schrell, and **M. Buchfelder**

Neurochirurgische Klinik, University of Erlangen-Nürnberg, Erlangen, Federal Republic of Germany

In 1929 Harvey Cushing defined the significance of the hypothalamus as follows: "Here in this well concealed spot almost all to be covered with a thumb-nail lays the very mainspring of primitive existence—vegetative, emotional, reproductive, on which with more or less success man has come to superimpose a cortex of inhibitions" (Cushing 1929). However it was Fröhlich who published the first hypothalamic syndrome in 1901 (Fröhlich 1901). Retrospectively his case appears to have been that of a cystic and haemorrhagic craniopharyngioma. Decades of research and late experience with hypothalamic hormones, developements of thin collimation computerized tomography (CT) and modern neurosurgical techniques in and around the hypothalamus have opened new doors. In 1983 clinicians and experts in basic research met in Munich to discuss modern knowledge of "clinical aspects of the hypothalamus" (Fahlbusch and Schrell 1985).

In spite of all the manifest complexity of pathophysiological mechanisms it is possible to distinguish between endocrinological and vegetative syndromes of the hypothalamus.

1. Endocrinological Syndromes

1.1 Influence of the Hypothalamus on Pituitary Adenomas

We are not able as yet to differentiate between possible hypothalamic and pituitary origins of central (ACTH-dependent) *Cushing's disease*. Nevertheless real recurrences in Cushing's disease do occur and possibly reflect a hypothalamic disturbance (Fahlbusch, Buchfelder and Müller 1986). In our series, we followed up 66 patients with complete clinical and endocrinological remission over an average period of 5.1 years. Nine of

them (12.1%) had true recurrences. The concept that central Cushing's disease could be induced by oversecretion of the hypothalamic corticotropin secreting neurons has not been proven by radioimmunological measurements of corticotropin releasing hormone itself (Stalla, Stalla, von Werder *et al.* 1987). Basal CRH plasma levels in remitting and non-remitting patients did not differ significantly between the two groups. In addition, the CRH test performed pre- and post-operatively gave no further information concerning the neurosurgical outcome. However in some patients the test gave indirect evidence for CRH deficiency and thus a pituitary origin of the disease (Schrell, Fahlbusch, Buchfelder *et al.* 1987). There is evidence that autoantibodies to pituitary ACTH cells play a significant role in patients whose Cushing's disease is refractory to treatment (Scherbaum, Schrell, Glück *et al.* 1987).

Hypothalamic releasing hormones may rarely cause hypersecretion of glandotrophic hormones. It has been shown that peripheral endocrine systems are able to synthesize peptides identical to the peptides produced by hypothalamic neurons. CRH has been found in the gut and the placenta (Stalla, Bost, Kaliebe *et al.* 1986) but tumours are also able to produce corticotropin (CRH) and growth hormone (GRH) releasing hormone. We have observed a 14-year-old giant, acromegalic girl with an *ectopic GRH-secreting* pancreatic tumour with liver metastases (von Werder, Losa, Müller *et al.* 1984). She had a suprasellar hyperplastic tumour with a chiasmal syndrome. After complete transsphenoidal tumour removal GH-levels decreased from a preoperative value of 450 ng/ml to 10 ng/ml in the early post-operative phase. However, GH increased within months up to 100 ng/ml documenting the unaltered GRH excess. Meanwhile both GRH and GH levels could be normalized by treatment with the somatostatin analogue SMS 201–995.

In prolactinomas recurrences occur in about 15% but the mechanism is not clear (Fahlbusch and Buchfelder 1985).

Hypersecretion of pituitary hormones can be caused also by the hypothalamic region itself, for example in true precocious puberty (Hadjilambris, Fahlbusch, Heinze, 1986). CT and nuclear magnetic resonance tomography (MRI) may demonstrate tumours developing from the tuber cinereum of the hypothalamus towards the suprasellar space. In cases which have a partial tumour stalk involvement, complete surgical removal with endocrinological remission is possible. In cases with a wide tumour connection to the hypothalamus no complete surgical removal is possible and symptomatic medical treatment with anti-androgens or a long-acting LHRH analogue is recommended (Comite, Pescovitz, Rieth *et al.* 1984).

Two girls have been successfully treated in our department. In both cases hamartomas were proven histologically but no LHRH immunoreactivity as reported by other authors (Price, Lee, Albright *et al.* 1984) could be demonstrated in the tumour tissue. From the pathological point of view there are two possibilities to explain LH-/FSH- or LRH-hypersecretion:

1. LH or LHRH could be secreted by the tumour itself. The peptide is packed in small dense granules, transported via axons to the median eminence or to the pituitary portal circulation and is released through fenestrated blood vessels as shown by electronmicroscopy.

2. The tumour may irritate the hypothalamus mechanically thus inducing LHRH hypersecretion.

In both our cases we could demonstrate that the pre-operative episodic secretion of LH and FSH in a range comparable to that of pubertal girls could no longer be demonstrated after successful surgery.

1.2 Endocrinological Disturbances Indicating Hypothalamic Damage. The Use of GRH- and CRH-Tests and the Insulin-Induced Hypoglycaemia Test

In normal persons the endogenous hypothalamic releasing hormones GRH and CRH stimulate the secretion of GH and ACTH (and subsequently cortisol) respectively. These glandotrophic hormones (GH and ACTH) can also be stimulated by the exogenous administration of the releasing hormones (GRH and CRH). Furthermore GH as well as ACTH and Cortisol can also be stimulated via the hypothalamus and the pituitary by the insulin induced hypoglycaemia (IH) test. When there is a disturbance of GRH and CRH production in the hypothalamus or its transport to the pituitary stalk, this stimulation of GH and ACTH release is deranged (Fahlbusch, Buchfelder and Schrell 1986).

This test is of real help in the early diagnosis of hypothalamic disturbances, at a stage when morphological changes of the hypothalamus are not easy to detect by CT or MR.

Case Example

We have observed an 18-year-old man who suffered from diabetes insipidus (DI) for 6 months. Endocrinological functional testing showed only non-tumourous hyperprolactinemia. Initially several thin collimation CT scans of the hypothalamic region could not demonstrate a pathological process. Endocrine function tests revealed that GH could be stimulated by GRH and cortisol and ACTH by CRH but not by insulin induced hypoglycaemia. Very careful examination of the CT showed an isodense area in the right suprasellar cistern, extenting from the hypothalamus. The germinoma was resected in part and disappeared after radiotherapy.

We have examined 44 other cases with tumours in the hypothalamic region. 15 had normal results; in 2 we found the typical abnormal regulation (described above) and in 7 of the remaining 27 cases it was not possible to differentiate between hypothalamic and pituitary stalk disturbance (Fahlbusch, Buchfelder and Schrell 1986). This test is helpful to diagnose an endocrine hypothalamic disturbance but there is no test that indicates the extent of morphological involvement of the hypothalamus.

2. Vegetative Syndromes

5 main vegetative disturbances may be distinguished:

1. Caloric balance: Obesity (Fröhlich's syndrome, 1901) and cachexia (Russel's syndrome, 1951).

2. Thermoregulation with hyper- and hypothermia (Clar 1985).

3. Sleep and wakefulness.

4. Electrolyte-water balance with hyper- and hyponatraemia, with and without polydipsia (Carmel 1980).

5. Miscellaneous:

a) Altered gastric physiology.

b) Diencephalic "epilepsy" (Penfield 1929): cutaneous flushing, hypertension, lacrimation, salivation, sweating, changes in the size of the pupils, Cheyne-Stokes respiration.

c) Laughing-seizures (Plouin, Ponsot, Dulac *et al.* 1983).

d) Mental changes *e.g.* anorexia, boulimia.

Case Example

We have seen a 20-year-old man with a chiasmal-syndrome and non-tumourous hyperprolactinaemia. He had the typical endocrinological constellation for hypothalamic defect. There was no increase of GH and cortisol after insulin hypoglycaemia but GH and cortisol could be stimulated by GRH and CRH. In addition the patient had hypernatraemia, and no diabetes insipidus. He had a germinoma of the third ventricle. After radiotherapy the hypernatreamia worsened and the patient was effectively treated with copious water input.

The role of vasopressin and natriuetic peptide requires eluciation (Lightman and Everitt 1986; see also this volume). Involvement of the hypothalamic centers for water and salt balance is still under discussion.

In the future, investigation of the clincal syndromes of the hypothalamus will have to focus on meticulous case examination including morphological, clinical, and laboratory data. It may be possible that MRI will provide by thin collimation of small hypothalamic areas, some morphological data to amplify Carmel's classification of the topographic relationship (anterior, preoptic, tuberal, and posterior hypothalamic region) to the special clinical syndromes (Carmel 1980).

In this way we can speculate with Carmel that contemporary immunohistochemical, electromicroscopic and microiontophoretic experiments have stressed its protean function, justifying the ancient idea that this small region is indeed "the site of the soul" (Carmel 1985).

References

Clar HE (1985) Disturbances of the hypothalamic thermoregulation. Acta Neurochir (Wien) 75: 106–112

Carmel PW (1980) Surgical syndromes of the hypothalamus. Clin Neurosurg 27: 133–159

— (1985) Tumours of the third ventricle. Acta Neurochir (Wien) 75: 136–146

Comite F, Pescovitz OH, Rieth KG, Dwyer K, McNemar A, Loriaux DL, Cutler jr GB (1984) Luteinizing hormone-releasing hormone analogue treatment of boys with hypothalamic harmatoma and true precocious puberty. J Clin Endocrinol Metab 59: 888–892

Cushing H (1929) The pituitary and hypothalamus. Charles C Thomas, Springfield, Ill

Fahlbusch R, Schrell U (1985) Surgical therapy of Lesions within the hypothalamic region. Acta Neurochir (Wien) 75: 125–135

— Buchfelder M (1985) Present status of neurosurgery in the treatment of prolactinomas. Neurosurg Rev 8: 195–205

— — Schrell U (1986) Endocrinological disturbances detected in hypothalamic lesions. In: Samii M (ed) Surgery in and around the brain stem and the third ventricle. Springer, Berlin Heidelberg New York, pp 367–374

— — Müller OA (1986) Transsphenoidal surgery for Cushing's disease. J Roy Soc Med 79: 262–269

Fröhlich A (1901) Ein Fall von Tumor der Hypophysis cerebri ohne Akromegalie. Wien Klin Rundschau 15: 883–906

Hadjilambris K, Fahlbusch R, Heinze E (1986) True precocious puberty of a girl with harmatoma of the CNS successfully treated by operation. Eur J Pediatr 145: 148–150

Lightman SL, Everitt BJ (1986) Water excretion. In: Lightmann SL, Everitt DJ (eds) Neuroendocrinology. Blackwell, Oxford, pp 197–206

Penfield W (1929) Diencephalic autonomic epilepsy. Arch Neurol Psychiat 22: 358–364

Plouin P, Ponsot G, Dulac O, Diebler C, Arthuis M (1983) Harmatomes hypothalamiques et crises de rire. Rev Électroencephalogr Neurophysiol Clin 13: 312–316

Price RA, Lee PA, Albright AL, Ronnekleiv OK, Gutai JP (1984) Treatment of Sexual Precocity by Removal of a Luteinizing Hormone-Releasing Hormone Secreting Harmatoma. JAMA 251: 17,2247–17,2249

Russell A (1951) A diencephalic syndrome emaciation in infancy and childhood. Arch Dis Child 26: 275

Scherbaum WA, Schrell U, Glück M, Fahlbusch R, Pfeiffer EF (1987) Autoantibodies to pituitary corticotropin-producing cells: Possible marker for unfavourable outcome after pituitary microsurgery for Cushing's disease. Lancet i: 1394–1398

Schrell U, Fahlbusch R, Buchfelder M, Riedl S, Stalla GK, Müller OA (1987) Corticotropin-releasing hormone stimulation test before and after transsphenoidal selective microadenomectomy in 30 patients with Cushing's disease. J Clin Endocrinol Metab 64/6: 1150–1159

Stalla GK, Bost H, Kaliebe T, Huber M, Stalla J, Pfeiffer D, Werder K von, Müller OA (1986) Human placental corticotropin releasing factor is identical to hypothalamic corticotropin releasing factor. Acta Endocrin (Kbh) 111 [Suppl 274]: 194

— Stalla J, Werder K von, Müller OA, Lüdecke PK, Schrell U, Fahlbusch R (1987) Corticotropin releasing hormone in plasma of patients with Cushing's disease. Klin Wochenschr 65: 529

Werder K von, Losa M, Müller OA, Schweiberer L, Fahlbusch R, Del Pozo E (1984) Treatment of metastasing GRF-producing tumour with a long acting somatostatin analogue. Lancet ii: 282–283

Acta Neurochirurgica, Suppl. 47, 61–67 (1990)

Pituitary Tumours: Problems and Questions

G. M. Teasdale[1], **A. M. McNicol**[2], **J. A. Thomson**[3], and **D. L. Davies**[4]

[1] University Department of Neurosurgery, Institute of Neurological Sciences, Southern General Hospital, Glasgow
[2] University Department of Pathology, Royal Infirmary, Glasgow
[3] University Department of Medicine, Royal Infirmary, Glasgow
[4] Department of Medicine, Western Infirmary, Glasgow, U.K.

Introduction

There have been many advances in the understanding and treatment of pituitary tumours in the last two decades. The basic scientific and clinical literature is now extensive and excellent reviews can be found in textbooks and journals. Despite this, important problems remain unresolved. This paper presents a personal viewpoint of these problems and seeks to identify some of the questions that neurosurgical research workers might aim to answer. These concern the methods of diagnosis of pituitary tumours and the assessment of their effects, their natural history, the indications for treatment and the advantages of different methods of management, how the results of treatment are assessed and how they may be improved.

Diagnosis and Initial Assessment

In each patient two questions need to be answered: first, does the patient have a pituitary adenoma or another disease of the sella region; and second, if an adenoma, of what specific kind? Advances in neuroradiology, biochemistry and endocrinology have made these questions more easy to answer, but are the answers they provide always correct? Pathological verification is always the final criterion but management decisions have to be taken before this is available.

Neuroradiologists now have a wide range of techniques that can be applied to the patient suspected of having a pituitary adenoma.

How much investigation is needed and, if unnecessary investigations should be minimized, what are the best strategies for different patients?

Cross-sectional imaging, either by CT or MRI, clearly has a central place in defining the location and anatomy of a sellar lesion. Nevertheless, a preliminary plain skull film remains useful; it can indicate the most appropriate technique for cross-sectional imaging (*e.g.* axial CT versus coronal CT); also, the finding of a pituitary fossa of normal size is a warning that what may appear on coronal imaging to be an intra-/suprasellar tumour, is a primary suprasellar meningioma. On the other hand, routine angiographic studies are probably no longer needed. The findings on plain film and either CT or MRI make it possible to select patients who need angiography to exclude an abnormal position of the carotid arteries, or a lesion such as an aneurysm (Macpherson *et al.* 1987). The relative merits of CT and MRI remain to be finally established. At present high resolution CT provides the information needed for the management of most patients. The advantages of MRI appear to lie mainly in better definition of the structures surrounding the tumour—especially the optic pathways and the CSF circulation, without the need for invasive methods. With both CT and MRI, the signals can provide some indication of the texture of the tumour (*e.g.* solid or cystic) but in neither case do they provide a specific histological diagnosis.

Biochemical studies have become important in diagnosis since recognition that some large pituitary tumours, as well as producing space occupying effects, also secrete a pituitary hormone, *e.g.* prolactin. These biochemical studies can influence management, in particular in determining if a "medical" approach offers a suitable alternative to operative treatment. Along with this has been an increasing tendency to base clinical reports upon the diagnosis finally arrived at, rather upon the patients' presenting problems. Unfortunately, the precise nature of a tumour may not be established until it has been studied in the laboratory, after removal. Retrospective reports can give a misleading impression about the pattern of tumour types seen in different clinical presentations. For example, reports from endocrine departments tend to over-emphasize the number of prolactin-secreting tumours, which may account for only a minority of cases in unselected series of large tumours.

There is still a need for reports of extensive series of patients, categorized according to their clinical problems but also fully investigated by modern techniques and managed by well-specified protocols. Such "problem-orientated" information provides a more appropriate basis for decisions about the management of newly presenting cases and for comparing the benefits of different approaches to management.

Classification of Pituitary Adenomas

The size and biological activity of a tumour determine the prognosis for the individual patient and provide an appropriate basis from which to classify patients in order to compare results of treatment in different series. Hardy's (1969) distinction between microadenoma (< 10 mm) and macroadenoma (> 10 mm) is widely accepted, but it has not been established how close a correspondence exists between radiological or operative measurements nor has the validity of either method been tested against the "gold standard" of pathology.

Existing classifications of tumour size and activity suffer from uncertainly about whether or not invasiveness can be assessed radiologically, at operation, or only after full histological examination (Balagura *et al.* 1981). There is debate about what degree of change in the radiological appearances of the pituitary fossa and its contents constitutes an unequivocal abnormality and about how to measure, and hence how to grade intra- and extrasellar abnormalities. The radiological classification of Hardy and Vezina (1976) has been found useful and has recently been revised. As well as validity, a system of classification needs to have high consistency; this is important in comparing results of treatment in different series. The inter-observer variability of systems for grading pituitary tumours needs to be established.

Choice of Management for a Large Tumour

Progressive visual failure remains the most common method of presentation of a large space occupying pituitary tumour. A transsphenoidal operation can be dramatically effective in improving vision and is employed by most surgeons whenever possible. At the same time, it is important to identify patients in whom transsphenoidal operation is not appropriate, either because the tumour is not a pituitary adenoma or because its configuration is not suitable.

Criteria for determining whether or not a tumour is suitable for transsphenoidal operation have not been established definitively. Surgeons in training often seek for guidelines in terms of absolute measurements, either of the size of the pituitary fossa or the suprasellar component. By contrast, experienced surgeons seem to base their judgements on a subjective assessment of the shape of the extrasellar parts of the tumour, their size in relation to the pituitary fossa and endeavour to use a transsphenoidal approach as often as possible. Sometimes this results in a less than complete removal of the suprasellar component of the tumour; despite this, vision may improve considerable. The lack of clear information about the relationship between the postoperative radiological appearances and the change in vision makes it difficult to determine if a particular degree of decompression is "adequate", if a persisting suprasellar remnant should be treated by a further transcranial operation or if it will "resolve" with time and after radiotherapy. In a series in Glasgow, recovery of vision was more rapid if the diaphragm collapsed into the fossa at the time of operation, but this did not influence significantly the eventual recovery (Findlay *et al.* 1983).

Medical Treatment of a Large Tumour

The likelihood that a large tumour will shrink satisfactorily in response to treatment, for example with bromocriptine, is an important factor in deciding about a patient's management.

If the patient's serum prolactin concentration is considerably raised it is very likely that the tumour is a prolactinoma, but the serum prolactin concentration can also be raised by other large tumours, as a result of interference with hypothalamic-pituitary control (Lees and Pickard 1987). To establish whether or not a tumour is a prolactinoma may need in vitro studies employing immunohistochemistry and assay of hormone production by tumour samples. These, of course, are not available when the clinical decision has to be

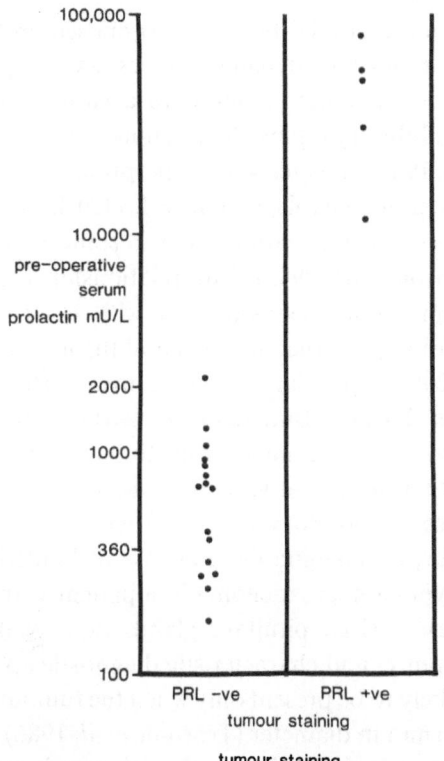

Fig. 1

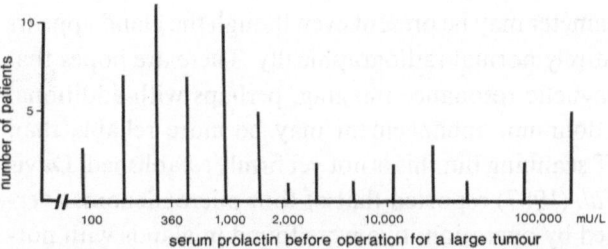

Fig. 2

made. Therefore, more correlations are needed between in vitro measurements and pre-operative endocrine studies. When prolactin is only moderately raised is it possible to distinguish reliably between secretion of true prolactinoma and the hyper-prolactinaemia that results from "compression" of hypothalamic pituitary perfusion?

The pre-operative serum prolactin found in a series of patients with a large tumour in Glasgow were compared with the results of post-operative immunohistochemistry and with results of studies of in vitro hormone production (Fig. 1). The results indicated two separate populations: patients whose tumour was composed of a large proportion of lactotrophs and was prolactin secreting in culture had a pre-operative serum prolactin concentration greater than 10,000 mu per li-

tre. In other cases the values were below 2,500 mu per litre.

The distribution of serum prolactin in a large series of patients, each of whom had presented with visual failure, also provided evidence for the existence of two subpopulations (Fig. 2). In one subgroup the prolactin concentrations were normal or only mildly raised, in the other there was extreme hyperprolactinaemia. On the other hand these data did not show a clear break point between these groups. The highest serum prolactin concentration reported in a patient whose tumour was diagnosed on the basis of immunocytochemistry not to be a prolactinoma is 6,000 mu/l. More data are needed about the likelihood that a tumour is a true prolactinoma or another type of adenoma, related to the patient's serum prolactin concentration in the range between 2,000 and 10,000 mu/litre.

Protocols Combining Different Treatment Methods

A patient with a prolactinoma may be treated either with bromocriptine or by operation. Usually one or other method is selected, with the second reserved until there is evidence that the first has not been completely effective. Alternative approaches have been described, in which the strategy is to precede operation with a course of treatment with bromocriptine. This may be given orally or in larger doses by a long-acting injection (Montini *et al.* 1986). The reduction in size during bromocriptine treatment may facilitate the subsequent surgical decompression, and this can be particularly helpful if the tumour has lateral or subfrontal extensions. Nevertheless, it seems unlikely that such a course of treatment will change whether or not a tumour can be excised so completely that recurrence does not occur, and it remains to be established if this approach improves upon the long-term outcome achieved by other strategies.

Numerous reports confirm that bromocriptine can shrink a true prolactinoma, but several aspects of its value remain uncertain (Lawton 1987). One problem is that most reports refer to treatment of fairly small series of highly selected patients, usually patients who did not have rapidly failing vision and who had only a visual field defect as opposed to loss of acuity. In the series of Molitch *et al.* (1985), of eight patients with impaired acuity, five improved but two appeared to worsen. Furthermore, not all prolactinomas become smaller with bromocriptine and when there is a response, the change may be relatively small and slow. The results of Scotti *et al.* (1982), Liuzzi, (1985) and

Molitch *et al.* (1985) indicate that some one third of patients fail to gain significant benefit and overall the results of bromocriptine do not appear to be as predictable or as rapid as those achieved by transsphenoidal decompression. Moreover, reports are beginning to appear of tumours initially regressing with bromocriptine treatment but subsequently recurring and progressing, despite continued therapy (Breidahl *et al.* 1983). What should be done when a tumour shows a distinct response to bromocriptine, but a considerable extrasellar mass remains after prolonged treatment? Some prolactinomas pursue an inexorable aggressive invasive course, despite operation, radiation, bromocriptine and other medical measures.

Diagnosis of a Hypersecreting Microadenoma

Diagnosis can be difficult in patients presenting with an endocrine disorder. Does the abnormality lie primarily in the pituitary or is there ectopic hormone production from another site in the body; also, is the pituitary abnormality a single discrete adenoma or a more diffuse or multifocal abnormality of pituitary structure and function, perhaps secondary to hypothalamic dysfunction? A further issue concerns the preoperative identification of the exact location of any tumour within the pituitary. Various endocrinological and radiological criteria are proposed for identifying a microadenoma but as yet none appear to be wholly reliable.

The need to distinguish between a pituitary and an ectopic source of hormone production is greatest in Cushing's syndrome (Howlett and Rees 1985). Recent additions to the available investigations are the study of the pituitary-adrenal response to corticotrophin releasing factor and the sampling of blood from the petrosal sinuses for determination of ACTH concentrations (Landolt 1986). The suggestion that comparison of the ACTH values obtained from the right and left petrosal sinuses can indicate the location of tumour in the pituitary needs more thorough evaluation (Oldfield *et al.* 1985). Many ACTH producing tumours lie in a central position in the gland and there are effective communications between the right and left cavernous sinuses.

Even when it seems clear that the pituitary is the source of inappropriate secretion of ACTH or prolactin, it would be valuable to be able to identify before operation which patients with Cushing's disease or hyperprolactinaemia have a discrete pituitary adenoma and which have diffuse pituitary dysfunction. A wide range of biochemical studies have been proposed and there are many reports of dynamic studies, assessing the response of basal hormone levels to a variety of stimulatory or inhibitory inputs. As yet, none is wholly reliable (Lancet 1987). The tests for hyperprolactinaemia appear to be more reliable than those for Cushing's disease but errors of classification occur in some 10% of cases (Thomson *et al.* 1985). The relationship between a microprolactinoma associated with clinical manifestations of hyperprolactinaemia and the lesions found in some 20% of pituitary glands at post mortem also needs to be clarified (Burrows *et al.* 1981). It has even been suggested that a microprolactinoma is not an abnormality but a normal variant (Lancet 1987)!

Each advance in radiology is greeted with enthusiasm but, so far, all imaging methods are of limited value in identifying a microadenoma in a patient with a relatively normal sized pituitary gland. Equivocal findings are common and characteristic diagnostic appearances are likely to be present only when the tumour is greater than 5 mm in diameter (Teasdale *et al.* 1986). Small focal changes in the radiodensity of the pituitary gland are equally likely to be a natural finding as a tumour (Brown 1983, Swartz *et al.* 1983, Rappoto and Latchaw 1983). Moreover, a microadenoma of < 5 mm diameter may be present even though the gland appears entirely normal radiographically. There are hopes that magnetic resonance imaging, perhaps with additional gadolinium enhancement may be more reliable than CT scanning but this is not yet firmly established. Davis *et al.* (1987) reported that of four microadenomas verified by operation, two were found in glands with normal MR appearances. Kucharczyk *et al.* (1986) reported finding a microadenoma in 10 of 11 patients who had normal MRI but, with proper caution commented "we recognize that our experience is drawn from a highly selected group of patients and hence the apparent sensitivity and specificity of MR imaging may be overstated".

What is the Natural History of a Pituitary Tumour?

The expected rate of progression is the yardstick against which the benefits of treatment of a pituitary tumour must be assessed. If space occupying effects already exist, it is reasonable to anticipate that these will persist and progress, but this may be at widely varying rates. There is even more uncertainty about what to expect when an adenoma is diagnosed before it has produced clinical effects. This may arise in the investigation of a patient presenting with pituitary insufficiency or as an incidental finding, for example after a head injury or in the investigation of "headache".

How often a microadenoma, and in particular a microprolactinoma, enlarges and at what rate is controversial. Weiss and colleagues reported that after six years observation there was evidence of growth in only 3 of 37 patients. Unfortunately, it is not clear if all 37 patients had a tumour; many had a relatively modest increase in serum prolactin concentration compatible with "functional" hyperprolactinaemia. Moreover, only 20 of the patients were followed for the full six years. Pontirolli and Falsetti (1984) found radiological evidence of tumour growth in a high proportion of patients considered to have a microprolactinoma, even though bromocriptine treatment had been given. It is uncertain if serial estimations of serum prolactin concentration can provide a reliable clue to tumour growth; examples of tumour progression without a change in serum prolactin were reported by Weiss *et al.* (1983).

The possibility of regrowth after decompression of a large tumour is the reason that radiotherapy is used frequently after operation. On the other hand, recurrence may be long delayed and perhaps not always inevitable; can patients be identified whose risk of recurrence is so low that radiotherapy is unnecessary as a routine? Histologically, pituitary adenomas show varying indices of growth potential but how reliably these predict recurrence has not been firmly established. Can newer methods of in vitro investigation provide more valuable biological information?

Morbidity and Mortality Resulting from Hormone Hypersecretion

Classical Cushing's disease is well established to have a poor prognosis and the benefits of treatment are clear (Hardy 1982, Fahlbusch 1986). Although less immediately life-threatening, the enhanced mortality resulting from acromegaly also is well established but there is less evidence about the degree to which life expectancy can be restored. There are well established correlations between the extent of growth hormone hypersecretion and the metabolic derangements present at the time of diagnosis (Davies *et al.* 1985). There is less clear information about the relation between the change in these measurements after operation and the prognosis for life expectancy. Should every acromegalic patient be offered operation, even very elderly patients with mild features? With hyper-prolactinaemia, as well as the risk of tumour growth and visual failure, the main concern long-term is the possibility of osteoporosis (Klibanski *et al.* 1980).

Results of Treatment

How often a "cure" can be achieved is unclear; indeed, it may be difficult to establish if "cure" has been accomplished. The problem is that after apparently complete removal of a tumour, microscopic remnants may be present in the remaining pituitary gland or in the dura of the walls of the pituitary fossa (Wrightson 1978). Such remnants may present without evidence of abnormal levels of hormone secretion, either under basal conditions or in response to stimulation.

There is a need to establish uniform criteria to assess the outcome of operation for a hypersecreting pituitary tumour. Measurements of serum hormone concentrations under basal conditions, and the construction of a diurnal profile from repeated sampling are useful indices of the severity of the persisting biochemical abnormality. The changes in pituitary secretion in response either to stimulation or to inhibition provide additional information, and may reveal abnormalities in patients with an apparently normal pattern of spontaneous secretion. Unfortunately, there is a lack of consistency about what basal values are regarded as "normal" and about the significance to attach to additional testing. Thus the serum prolactin concentration regarded as normal ranges from 250 to 1,000 mu/l. The trend with acromegaly as experience increases is for criteria to become increasingly rigorous. Thus the values of serum growth hormone concentration (either a basal level or as a mean of a diurnal profile) considered to be "normal" have fallen from 20 IU/litre down to 5 IU/litre (Schaison *et al.* 1983). The latter value is supported by the results of a comparison in Glasgow of post treatment serum growth hormone and metabolic indices (McLellan 1988). There is a clear need to establish valid and uniform criteria.

Persistent or Recurrent Hypersecretion After Operation: Improving Results

The persistence of hypersecretion following operation for a hormonally active adenoma is well known. There is now greater interest in patients who, after an apparently successful operation, subsequently suffer recurrent hypersecretion. Do such cases reflect pituitary hypersecretion due not to a single adenoma but due either to multifocal lesions or to a diffuse disorder? Is recurrence a response to a primary hypothalamic defect? Alternatively, do such cases simply reflect a failure to remove completely a single adenoma, so that remnants persist in the remaining pituitary gland or in the dura around the fossa? Answers also are needed to such questions as: does pre-operative treatment of a prolactinoma with bromocriptine improve or worsen the post-operative results; can results be improved by modifications in operative technique, for example by radically excising apparently normal pituitary tissue as well as the tumour and its capsule or by per-operative monitoring of hormone levels; is there a benefit in treating the walls of the cavity of the tumour with alcohol or another anti-tumour agent; in which patients is a further operation, with greater risks (Laws 1985) indicated and which patients need post-operative radiotherapy and how valuable and how safe is this?

Reported rates of recurrence after operations for acromegaly (Serri *et al.* 1985), Cushing's disease (Busch 1983) or hyperprolactinaemia vary widely. A report of 55% recurrence of hyperprolactinaemia after operation for prolactinoma (Serri *et al.* 1983) has not been substantiated by subsequent studies which have shown a much lower rate of recurrence (Thomson *et al.* 1985). Some of the differences can be explained by the varying criteria used to decide if the initial response to surgery was adequate and by the varying evidence used to determine that subsequent recurrence has occurred. Nevertheless, does some of the variation indeed reflect differences in surgical method? The role of the vasculature of a pituitary tumour needs more study (Powell *et al.* 1974, Erroi *et al.* 1986). There is reason to believe that the sources of excessive secretion in hyperprolactinaemia are lactotrophs perfused not by hypothalamic-pituitary portal blood but from branches of the intracavernous carotid artery (Teasdale 1983, Elias and Weiner 1984). Is such vascularization merely a secondary consequence of tumour growth or does it have a primary role either in tumour development and progression, or in the late recurrence of hyperprolactinaemia?

Laboratory Research

The ability to identify tumour cell types by immunocytochemistry (Kovacs 1981) electronmicroscopy (Landolt 1984) and to assay hormone production from pituitary tumour samples (Adams and Mashiter 1985, Landolt *et al.* 1986, Stefanko *et al.* 1983) have provided fruitful bases for research. They have led to much more refined and relevant methods of classification and to much knowledge of the mechanisms of control of secretion from adenoma cells. However, such studies have had (as yet?) little influence on the management of a new patient with an adenoma.

Advances in the management of pituitary tumours may depend upon the application of modern methods for studying the mechanisms involved in tumour growth and viability. The neurosurgon wishing to improve understanding and treatment of pituitary tumours may, in the future, need more familiarity with the techniques of molecular pathology than knowledge of the pharmacological and biochemical methods that have until now been the mainstay of much of neuroendocrinological research.

References

Adams EF, Mashiter K (1985) Role of cell and explant culture in the diagnosis and characterization of human pituitary tumours. Neurosurg Review 8: 135–140

Balagura S, Derome P, Guiot G (1981) Acromegaly analysis of 132 cases treated surgically. Microsurgery 8: 413–416

Breidahl H, Topliss JD, Pike JW (1983) Failure of bromocriptine to maintain reduction in size of a macroprolactinoma. Br Med J 287: 451–452

Brown SB, Murphy Irwin K, Enzmann DR (1983) CT characteristics of the normal pituitary gland. Neuroradiology 24: 259–262

Burch W (1983) A survey of results with transsphenoidal surgery in Cushing's disease. N Engl J Med 308: 103–104

Burrow GN, Wortzman G, Rewcastle NB, Holgate RC, Kovacs K (1981) Microadenomas of the pituitary and abnormal sella tomograms in an unselected autopsy series. New Engl J Med 304: 156–158

Davies DL, Beastall GH, McConnell J, Fraser R, McCruden D, Teasdale GH (1985) Body composition, blood pressure and the renin angiotensin system in acromegaly before and after treatment. J Hypertension 3: S 413–S 415

Davis PL, Hoffman JC, Spencer T, Tindall GT, Braun IF (1987) MR imaging of pituitary adenomas: CT, clinical and surgical correlations. AJNR 8: 107–110

Elias KA, Weiner RJ (1984) Direct arterial vascularization of estrogen induced prolactin-secreting anterior pituitary tumours. Proc Natl Acad Sci USA 81: 4549–4553

Erroi A, Bassetti M, Spada A, Giannattasio G (1986) Microvasculature of human micro- and macroprolactiomas. Neuroendocrinology 43: 159–165

Fahlbusch R, Buchfelder M, Muller OA (1986) Transsphenoidal surgery for Cushing's syndrome. J Roy Soc Med 79: 202–209

— — Schell V (1987) Short term pre-treatment of macroprolactinomas by dopamine agonists. J Neurosurg 67: 807–815

Findlay G, McFadzean RM, Teasdale GM (1983) Recovery of vision following treatment of pituitary tumours, application of a new system of assessment to patients treated by transsphenoidal operation. Acta Neurochir 68: 175–186

Hardy J, Vezina JL (1976) Transsphenoidal neurosurgery of intracranial neoplasms. In: Thompson RA, Green JR (eds) Advances in neurology. Raven Press, New York, pp 761–774

— (1982) Cushing's disease 50 years later on. Can J Neurol Sci 9: 375–380

— (1969) Transsphenoidal microsurgery of the normal and pathological pituitary. Clin Neurosurg 10: 185–217

Howlett TA, Rees LH (1985) Is it possible to diagnose pituitary-dependent Cushing's disease? Ann Clin Biochem 22: 550–558

Klibanski A, Neer RM, Beitins IZ, Ridgway EC, Zervas NT, McArthur JW (1980) Decreased bone density in hyperprolactinaemic women. New Engl J Med 303: 1511–1514

Kovacs K, Horvath E (1986) Tumours of the pituitary gland. Armed Forces Institute Pathology Publications, Washington D.C., 269 p

Kucharcyk W, Davis DO, Kelly WM, Sze G, Normal D, Newton TH (1986) Pituitary adenomas: high resolution MR imaging at 1.5T. Radiology 161: 761–765

Lancet editorial (1987) Hyperprolactinaemia: when is a prolactinoma not a prolactinoma? Lancet ii: 1002–1004

Landolt AH (1984) Secreting pituitary adenomas: morphological and biological aspects in pituitary hyperfunction. In: Camanni F, Miller, E (eds) Serino symposia publications vol 10. Raven press, New York, pp 135–155

Landolt AM, Hertz PV (1986) Alpha-Subunit-producing pituitary adenomas: immunocytochemical and ultrastructural studies. Virchows Arch (Pathol Anat) 409: 417–431

— Valavanis A, Girard J, Eberle AN (1986) Corticotrophin releasing factor test used with bilateral simultaneous inferior petrosal sinus blood-sampling for the diagnosis of pituitary dependent Cushing's disease. Clin Endocrinol 14: 133–143

Laws ER, Fode NL, Redmond MJ (1985) Transsphenoidal surgery following unsuccessful prior therapy. J Neurosurg 63: 823–829

Lawton NF (1987) Prolactinomas: medical or surgical treatment? Quart J Med 64: 557–564

Lees P, Pickard JD (1987) Hyperprolactinaemia intrasellar pituitary tissue pressure and the pituitary stalk compression syndrome. J Neurosurg 67: 192–196

Liuzzi A, Dallabonzana D, Opizzi G et al. (1985) Low doses of dopamine agonists in the long-term treatment of macroprolactinoma. New Engl J Med 313: 656–659

Macpherson P, Teasdale E, Hadley DM, Teasdale G (1987) Invasive v non-invasive assessment of the carotid arteries prior to transsphenoidal surgery. Neuroradiology 29: 457–461

McLellan A, Connell JMC, Beastall J et al. (1988) Growth hormone, body composition and somatomedin C after treatment of acromegaly. Quart J Med 69: 997–1008

Molitch ME, Elton RL, Blackwell RE et al. (1985) Bromocriptine as a primary therapy for prolactin-secreting macroadenomas— results of a prospective multi-center study. J Clin Endocrinol Metab 60: 698–705

Montini M, Pagani G, Goanola D et al. (1986) Long lasting suppression of prolactin secretion and rapid shrinkage of prolactinomas of a long acting injectable form of bromocriptine. J Clin Endocrinol Metab 63: 266–268

Oldfield EH, Chrousos GP, Schultz HM, Schaaf M, McKeevert PE, Krudy AG, Cutter GB, Loriauz DL, Doppman FL (1985) Preoperative lateralization of ACTH-secreting microadenoma by bilateral and simultaneous inferior petrosal sinus sampling. New Engl J Med 312: 100–103

Pontirolli AE, Falsetti L (1984) Development of pituitary adenoma in women with hyperprolactinaemia: clinical, endocrine and radiological characteristic. Br Med J 288: 515–518

Powell DF, Baker HZ, Laws ER (1974) The primary angiographic findings in pituitary adenomas. Radiography 110: 589–595

Rappolo HM, Latchaw RE (1983) Normal pituitary gland 1) Macroscopic anatomy—CT correlation. AJNR 4: 927–936

Schaison G, Louzinet D, Moatti N, Pertuiset B (1983) Critical study of the growth hormone response to dynamic tests and the insulin growth factor assay in acromegaly after microsurgery. Clin Endocrinol 14: 133–143

Scotti G, Scialfa G, Pieralli S, Chiodini PS, Spelta B, Dallabonzana D (1982) Macroprolactinomas: CT evaluation of reduction of tumour size after medical treatment. Neuroradiology 23: 123–126

Serri O, Rasio E, Beauregard H, Hardy J, Somma M (1983) Recurrence of hyperprolactinaemia after selective transsphenoidal adenomectomy in women with prolactinoma. New Engl J Med 209: 280–283

Stefanko SZ, Lamberts SWJ, Osteram R (1983) Relationship between cytological picture and hormone secretion by human pituitary adenomas in cell culture. Arch Neuropathol (Berl) OL 153–156

Swartz JD, Russell K, Basile BA, O'Donnel PC, Popky GL (1983) High resolution computed tomographic appearance of the intrasellar contents in women of child-bearing age. Radiology 147: 115–117

Teasdale G (1983) Surgical management of pituitary adenoma. Clin Endocrinol Metab 12: 789–823

Teasdale E, Teasdale GM, Mohsen F, Macphersen P (1986) High resolution computed tomography in pituitary microadenomas: is seeing receiving? Clinical radiology 37: 227–232

Thomson JA, Teasdale GM, Gordon D, McCruden D, Davies DL (1985) Treatment of presumed prolactinoma by transsphenoidal operation; early and late results. Br Med J 29: 1550–1553

Weiss MH, Teal J, Gott P (1983) Natural history of microprolactinoma, six year follow up. Neurosurgery 12: 180–183

Wrighton P (1978) Conservative removal of small pituitary tumours: is it justified by the pathological findings? J Neurol Neurosurg Psychiat 41: 183–299

Correspondence: G. M. Teasdale, University Department of Neurosurgery, Institute of Neurological Sciences, Southern General Hospital, Glasgow G51 4TF, U.K.

Acta Neurochirurgica, Suppl. 47, 68–70 (1990)

Intrasellar Pressure

P. D. Lees

Wessex Neurological Centre, Southampton General Hospital, Shirley, Southampton, U.K.

I. Introduction

The adenohypophysis receives its blood supply from two portal venous systems through fenestrated venules which are structurally similar to capillaries (Page 1986). The orthodox view that there is no direct arterial supply has recently been challenged (Gorczyca and Hardy 1987) but it is accepted that any arterial supply only serves a small minority of the gland. This unusual anatomy suggests a fundamental difference in the dynamics of adenohypophysial perfusion when compared to other organs in the body and assumes particular importance because of the dual nutritive and neuroendocrine functions of the pituitary portal blood supply.

Perfusion Pressure

The perfusion pressure of the brain (CPP) depends upon the difference between the vascular input (mean arterial) and intracranial pressures:

$$CPP = MAP - ICP$$

(MAP = Mean Arterial Pressure; ICP = Intracranial Pressure; CPP = Cerebral Perfusion Pressure)

Similarly, adenohypophysial perfusion pressure may be derived but the input pressure is portal venous pressure:

$$APP = PVP - ISP$$

(PVP = Portal Venous Pressure; ISP = Intrasellar Pressure; APP = Adenohypophysial Perfusion Pressure)

Portal Venous Pressure

The structure of pituitary portal veins is not designed to withstand high intraluminal pressures. Portal venous pressure is unknown but is likely to approximate to 10–15 mmHg, the venous pressure at the distal end of a standard capillary plexus. Thus, the vascular input pressure to the adenohypophysis is considerably lower than the input pressure to, for example, the brain.

Intrasellar Pressure

The normal pituitary gland is enclosed by the pituitary fossa and, therefore, expansion of the intrasellar contents by, for example the growth of a tumour, is likely to cause a rise in intrasellar pressure in the same way that brain tumours cause a rise in intracranial pressure. Normal intrasellar pressure is unknown but it can be argued that it does not exceed normal intracranial pressure (less than 10 mmHg). The two compartments are separated by the diaphragma sellae, the position of which may be considered a crude monitor of the relative compartmental pressures. In the normal subject, the diaphragma is flat; if deficient, the arachnoid prolapses into the sella giving rise to the empty sella syndrome (de Divitiis et al. 1981). In the empty sella syndrome, intrasellar and intracranial pressure must be equal. This does not disturb pituitary function in the majority and presumably, therefore, does not significantly affect adenohypophysial perfusion (Neelon et al. 1973).

Adenohypophysial Perfusion Pressure

Normal Physiology

The theoretical values of intrasellar pressure and portal venous pressure suggest that normal adenohypophysial perfusion pressure is below 10 mmHg. Consequently, adenohypophysial perfusion must be particularly vulnerable to relatively small changes in portal venous pressure or intrasellar pressure. Corroborative evidence for this hypothesis is found in the study of Antunes *et al.* (1983) who observed portal venous flow in the rhesus monkey by a Doppler technique. Complete arrest of long portal flow was produced by a Valsalva manoeuvre in which the positive airways pressure was only 30 cm H_2O (22 mmHg). It is suggested that the mechanism of this observation is that raised intrathoracic pressure was transmitted to the pituitary fossa by the venous system causing an increase in intrasellar pressure sufficient to exceed portal venous pressure. This is indirect evidence to suggest that normal portal venous pressure in the rhesus monkey is below 22 mmHg; the pituitary vascular anatomy of this primate is very similar to man (Carmel *et al.* 1979, Daniel and Prichard 1975, Antunes *et al.* 1983).

Pathophysiology

The apparently precarious nature of adenohypophysial perfusion raises speculation concerning the effects of the development of a pituitary tumour which may, as discussed above, cause a rise in intrasellar pressure.

Pituitary portal venous obstruction by raised intrasellar pressure is likely to have neuroendocrine sequelae: because the predominant nature of adenohypophysial control is stimulatory, hypopituitarism is the likely consequence. However, prolactin control is predominantly inhibitory and hence hyper- and not hyposecretion of this hormone is likely. Both hypopituitarism and hyperprolactinaemia are seen in patients with pituitary tumours and the latter is not always associated with a prolactin-secreting tumour. In the Stalk Compression Syndrome (SCS), moderate hyperprolactinaemia complicates usually a non-secreting tumour of the gland (Nabarro 1982) and, whilst often ascribed to portal venous obstruction, the suggestion that this is due to raised intrasellar pressure has not previously been made.

To explore this theory further, adenohypophysial perfusion pressure has been examined. It was felt that the direct measurement of portal venous pressure is impossible due to the delicate structure and anatomical position of portal veins but a technique has been devised for the measurement of intrasellar pressure (Lees and Pickard 1987). The method is invasive and has, therefore, only been applied to pituitary tumour patients at the time of operation.

II. Intrasellar Pressure Measurement

Intrasellar pressure has been measured at the time of transsphenoidal surgery in a series of patients with a variety of pituitary tumours. The mean age at the time

of surgery for the group of 33 patients was 43 years; there were 17 men and 16 women. Thirteen patients had non-secreting and twenty had endocrine-active tumours. The original series showed no correlation between hormone secretory pattern and intrasellar pressure so this was not examined in this study. Seventeen tumours had suprasellar extension demonstrated by computerised tomography.

The technique has been described in detail previously (Lees and Pickard 1987). In summary, pressure is measured through a fine needle which is inserted into the pituitary fossa through a small drill hole prior to full decompression of the fossa floor. The updated results of the first 33 patients are given here.

III. Results

Waveform

A characteristic waveform was seen in the majority of patients but was difficult to obtain in either very small or very firm tumours. The wave ocurred in synchrony with the arterial pressure wave which it partly resembled. A variable amplitude was seen with a maximum of 12 mmHg; the mean pulse pressure was 4 ± 0.5 mmHg (SEM) for the group as a whole.

The recorded mean intrasellar pressure varied from 2 to 51 mmHg with a mean for the group of 24 ± 2 mmHg. In more than 40% of patients, intrasellar pressure exceeded 30 mmHg.

Endocrinology

Pituitary function was compared to intrasellar pressure: Table 1 compares normal function with hypopituitarism, the latter being defined as a deficiency in one or more sector of pituitary function.

Because of its different (inhibitory) mechanism of control (*vide supra*), prolactin levels were treated separately (Table 2). The distinction between stalk compression syndrome and prolactinoma was based either upon the *in vitro* perfusion behaviour of tumour tissue (Lawton *et al.* 1981) or upon the serum prolactin level;

Table 1. *Correlation between preoperative pituitary function and intrasellar pressure**

	n	ISP ± SEM (mmHg)
Normal pituitary function	9	16 ± 3
Hypopituitarism	15	27 ± 3
P < 0.05 (analysis of variance)		

* Figures refer to original sizes.

Table 2. *Serum prolactin*

	n	ISP ± SEM (mmHg)
Normal	7	25 ± 3
Stalk compression syndrome	14	28 ± 3* **
Prolactinoma	7	17 ± 5**
On bromocriptine	5	14 ± 3*

* P < 0.01.
** P < 0.05.

a level of > 2,000 mIU/Litre was taken to imply a prolactinoma (Lawton *et al.* 1986).

Intrasellar pressure was highest in the stalk compression syndrome group and significantly lower in the prolactinoma group (P < 0.05). There was, however, little difference between intrasellar pressure in patients with normal prolactin levels (25 ± 3 mmHg) and those with stalk compression syndrome (28 ± 3 mmHg).

IV. Discussion

It has been argued that the adenohypophysial blood supply is delivered at unusually low pressure and is, therefore, particularly vulnerable to local tissue pressure changes which have not previously been quantified. This study shows that intrasellar pressure in pituitary tumours often exceeds the theoretical norm for portal venous pressure. Furthermore, the expected effects of portal venous obstruction, namely stalk compression syndrome and hypopituitarism, were found to be associated with siginficantly higher intrasellar pressures than other patients. This supports the hypothesis that endocrine dysfunction associated with pituitary tumours may in part reflect disturbances in portal blood flow.

The generation of higher levels of intrasellar pressure and the survival of the adenohypophysis denied its portal blood supply are, it is suggested, the consequences of neovascularisation involving the development of a collateral arterial blood supply (Baker 1971, Elias and Weiner 1984, Lees and Pickard 1987).

The arguments proposed in this study contain much supposition because of the lack of previously published data but the basic hypothesis that adenohypophysial perfusion pressure is very low seems to be borne out by the two studies in which it is considered. However, many questions are raised: what are the implications of a low perfusion pressure for normal adenohypophyseal function? How is normal adenohypophysial viability maintained if its portal blood supply is cut off? Is there a direct arterial blood supply to the normal adenohypophysis as suggested by McConnell (1953)

and Gorczyka and Hardy (1987) and denied by Daniel and Prichard (1975)? Why is adenohypophysial function normal in some patients with a high intrasellar pressure whilst others with similar pressures develop hypopituitarism and/or the stalk compression syndrome.

The questions thus exceed the answers! However, it is hoped that this study will provide a stimulus to further research into a new approach to neuroendocrine pathophysiology.

References

Antunes JL, Muraszko K, Stark R, Chen R (1983) Pituitary portal blood flow in primates: A Doppler study. Neurosurgery 12: 492–495

Baker HL (1972) The angiographic delineation of sellar and parasellar masses. Radiology 104: 67–78

Carmel PW, Antunes JL, Ferin M (1979) Collection of blood from the pituitary stalk and portal veins in monkeys, and from the pituitary sinusoidal system of monkey and man. Neurosurg 50: 75–80

Daniel PM, Prichard MML (1975) Studies of the hypothalamus and the pituitary gland. Acta Endocrinol (Copenh) 80 [Suppl 201]: 1–216

de Divitiis E, Spaziante R, Stella L (1981) Empty sella and benign intrasellar cysts. In: Krayenbühl H *et al.* (eds) Advances and Technical Standards in Neurosurgery, Vol 8. Springer, Wien New York, pp 1–75

Elias KA, Weiner RI (1984) Direct arterial vascularization of estrogen-induced prolactin-secreting anterior pituitary tumors. Proc Nat Acad Sci USA 81: 4549–4553

Gorczyca W, Hardy J (1987) Arterial supply of the human anterior pituitary gland. Neurosurgery 20: 369–378

Lawton NF, Evans AJ, Weller RO (1981) Dopaminergic inhibition of growth hormone and prolactin release during continuous *in vitro* perifusion of normal and adenomatous human pituitary. J Neurol Sci 49: 229–239

— — Pickard JD, Perry S, Davies B (1986) Secretion of neurone-specific enolase, prolactin, growth hormone, luteinising hormone and follicle stimulating hormone by "functionless" and endocrine-active pituitay tumours *in vitro*. J Neurol Neurosurg Psychiatry 49: 574–580

Lees PD, Pickard JD (1987) Hyperprolactinemia, intrasellar pituitary tissue pressure and the pituitary stalk compression syndrome. J Neurosurg 67: 192–196

McConnell EM (1953) The arterial blood supply of the human hypophysis cerebri. Anat Rec 115: 175–201

Nabarro JDN (1982) Pituitary prolactinomas. Clin Endocrinol 17: 129–155

Neelon FA, Goree JA, Lebovitz HE (1973) The primary empty sella: clinical and radiographic characteristics and endocrine function. Medicine, Baltimore 52: 73–92

Page RB (1986) The pituitary portal system. In: Ganten D, Pfaff D (eds) Current topics in neuroendocrinology, Vol 7. Springer, Berlin Heidelberg New York, pp 1–47

Correspondence: P. D. Lees, F.R.C.S., Wessex Neurological Centre, Southampton General Hospital, Shirley, Southampton S09 4XY, U.K.

Acta Neurochirurgica, Suppl. 47, 71–85 (1990)

The Medical Treatment of Prolactin and Growth Hormone-Secreting Pituitary Tumours

Ioana Lancranjan

Department of Neuroendocrinology, Clinical Research, Sandoz Ltd., Basle, Switzerland

Introduction

Prolactin (PRL) and growth hormone (GH)-secreting pituitary tumours are the most frequently occurring, diagnosed and treated pituitary tumours[49]. In a series of 424 cases of secreting pituitary tumours Hardy found 225 PRL-secreting adenomas and 199 GH-secreting adenomas[46]. The high incidence of PRL and GH-secreting pituitary tumours has been reported by many other groups[31, 50, 68, 72].

Important progress in biochemical and imaging techniques in recent years has provided a more accurate diagnosis of pituitary tumours. Sophisticated morphological techniques and receptor studies applied to surgical specimens has led to a better understanding of pituitary cytopathology and a better knowledge and treatment of pituitary tumours.

Classically, pituitary tumours were differentiated histologically as chromophobe, acidophil (eosinophilic) or basophil, depending on their staining characteristics with hematoxylin and eosin. With attempts to correlate the morphological features of the tumour cells with their secretory activity, clinical history, symptomatology and biochemical findings, Kovacs and Horvath[1] classified pituitary adenomas into eight subgroups: 1) PRL cell adenoma; 2) GH cell adenoma; 3) mixed GH and PRL cells adenoma; 4) acidophil stem cell adenoma; 5) corticotroph cell adenoma; 6) thyrotroph cell adenoma; 7) gonadotroph cell adenoma; and 8) undifferentiated cell adenoma, including oncocytoma.

The purpose of the present paper is not to argue the relative merits of the medical and surgical treatment but to briefly review the progress made in the last decade in the medical management of the first three types of pituitary tumours in the classification of Kovacs and Horvarth, with a special emphasis on new approaches to the treatment of PRL- and GH-secreting adenomas.

The Medical Approach to the Management of Prolactinomas

The objective of the treatment of a patient with prolactinoma is to lower PRL plasma levels to within the normal range, to restore gonadal function and other anterior pituitary hormones secretion to normal levels and to either remove or reduce the size of the tumour.

Two major approaches to the treatment of prolactinomas were developed in parallel in the last decade: 1) selective, transsphenoidal surgical removal of the tumour and 2) medical therapy with dopamine agonists.

This chapter will review the medical therapy of prolactinomas with respect to 1) rationale for the use of dopamine agonists in the management of prolactinomas; 2) medical therapy with bromocriptine of microprolactinomas and 3) macroprolactinomas; 4) other drugs, besides bromocriptine, used in the treatment of prolactinomas; and 5) new galenical forms of bromocriptine in the initial and/or long-term treatment of patients with prolactinomas.

Rationale for the Use of Dopamine Agonists in the Management of Prolactinomas

Physiologically the most important neural control of PRL secretion is exerted by dopamine (DA) which is secreted by the tuberoinfundibular neurons with cell bodies in the arcuate nucleus and with nerve endings on portal vessels in the median eminence. DA is the major PRL inhibiting factor in all mammals. DA agonist drugs bind to DA receptors on the lactotroph cells and lower PRL secretion in normal and pathological conditions, such as functional and tumoural hyperprolactinemia.

Bromocriptine, the first dopamine D_2 agonist used therapeutically in the treatment of hyperprolactinemic states, has been shown to produce both the suppression of PRL secretion and significant shrinkage of PRL-secreting tumours. During the first days of treatment, the shrinkage results from the involution of adenomatous PRL cells with reduction in cytoplasmic, nuclear and nucleolar areas[5, 88]. After long-term treatment with bromocriptine the histological findings show widespread necrotic areas in the tumours[41, 81] suggesting that bromocriptine may have both cytocidal and cytostatic effects on prolactinomas.

Because of the rapid improvement of clinical symptoms/signs of hyperprolactinemia and the effectiveness of bromocriptine in reducing tumour size, medical treatment has become, in many centres, the management of choice of prolactinomas.

Papers describing a high recurrence rate in surgically treated patients[82, 86] with prolactinomas have not been contained in present series[2, 16, 33, 43, 47, 79, 91, 97], but have suggested that some patients may have an underlying abnormality in the pituitary or hypothalamus which is not amenable to surgical cure.

The development of transphenoidal microsurgical selective adenomectomy marked a great step forward in the management of prolactinomas and high cure rates were reported[1, 27, 30, 45, 57, 62, 63, 87, 90]. Some of these reports considered cured the patients with normal PRL plasma levels immediately after surgery and may have overestimated the success of surgical treatment of prolactinomas.

Frantz et al.[39] found that the PRL plasma concentrations measured one week after surgery were significantly lower than that measured one month after surgery. Moreover, Serri et al.[82], in a series of 44 women followed for an average of 6.2 ± 1.5 years showed that relapse of hyperprolactinemia usually occurred one year after surgery and that the risk increases with time. Recurrence of hyperprolactinemia was observed by Serri et al. in 12 of 24 (50%) of patients with microprolactinomas and in $^4/_5$ (80%) of patients with macroprolactinomas. In the series reported by Rodman et al.[79] and others the incidence of recurrence is much lower. Rodman et al.[79] reported relapse in 17% of the 88% patients with microadenomas cured by surgery

and in 20% of the 37% of patients with macroadenomas with normal PRL secretion after surgery. Prolactin plasma levels were significantly higher both immediately and six weeks after surgery in patients who relapsed than in those patients who did not relapse[11, 13, 14].

Examination of PRL secretory dynamics, assessed by TRH and insulin-induced hypoglycaemia, six weeks postoperatively did not predict relapse of hyperprolactinemia[79]. Abnormalities of the dynamic test with domperidone showed a good correlation with late recurrence of hyperprolactinemia in patients with normal PRL plasma concentrations after surgical treatment of prolactinomas[17].

The theoretical advantage of surgical treatment of prolactinomas is that once completed successfully the patient may be cured for life. However, even the best surgical results compare favourably with medical therapy only for microprolactinomas whereas the surgical success in curing patients with large prolactinomas ranges from 0–30%[10, 63, 87, 90, 98]. The recurrence rate of hyperprolactinemia (17–80%) after apparently successful surgery and the undefined mechanism of tumour such relapse (side supra) has contributed to bromocriptine being considered as primary not just adjuvant treatment for both small and large prolactinoma.

Medical Treatment of Microprolactinomas: Effects of Bromocriptine

Essentially, PRL secretion is lowered to within the normal range in most patients treated with bromocriptine irrespective of the pretreatment PRL levels[10, 98]. Restoration of normal gonadal function and fertility occurs in the majority of patients including some where the radioimmunoassay of PRL plasma levels remained elevated above the normal range[85]. Bergh and Nillius[7] reviewed the results of bromocriptine therapy in 160 patients, $^{101}/_{160}$ with pituitary tumours, and found that the overall ovulation rate was 93% and the pregnancy rate 88%. In most women normal menses are resumed within 4 weeks after the start of treatment, although in some patients a considerably longer period of treatment is necessary to restore regular cycles[8, 40]. The discontinuation of bromocriptine therapy is followed in most cases by a rise of PRL plasma levels to either pretreatment or slightly lower levels. Isolated cases with persistent normalization of PRL levels were however reported[28, 29, 100]. Faglia et al.[29] reported persistent normalization of PRL plasma levels in $^4/_{21}$ patients who underwent long-term treatment with bromocriptine. In another paper Faglia's group reported clinical remission and normal PRL levels in $^4/_{36}$ women with microprolactinomas treated with bromocriptine (2.5–10 mg/day) for 12 months and in $^4/_{18}$ women who extended the therapy for another 12 months[67].

Tumour regression or disappearance of microprolactinomas documented by high resolution CT scan examinations were also reported[12, 26, 29]. Tramu et al.[89] reported marked morphological changes in four microadenomas surgically removed after 6–17 months therapy with bromocriptine. In 3 cases in whom PRL secretion was normalized by bromocriptine therapy, fibrosis, pyknotic cells and crinophagic appearances were found in the adenomas. In one case total necrosis and fibrosis of the adenoma was found.

Fossati and Mazzuca[38] reported a well documented case of microadenoma cured after 30 months' treatment with bromocriptine 7.5 mg daily. The patient was a 28-year-old man with hyperprolactinemia, bilateral gynecomastia, galactorrhea and decreased libido. The basal PRL level was 205 ng/ml and the sella turcica exhibited a double contour of the floor with localized rupture of the anterio-inferior curve. After bromocriptine withdrawal, which followed the 30 months treatment period, the man was clinically normal, plasma PRL levels were in the normal range and the sella turcica was thickened with recalcification of its floor for the whole follow-up period of 18 months.

Pregnancy occurring during bromocriptine therapy has a normal evolution and ends with a normal delivery of a normal child[92, 96]. Enlargement of the pituitary tumour during pregnancy occurs very seldom[7, 96]. Remission of hyperprolactinemia after bromocriptine-facilitated pregnancy has been reported by several groups[96] after the first report published by Cowden and Thomsen[22]. Their patient had tomographic evidence of a pituitary tumour in whom bromocriptine facilitated pregnancy was followed by breast-feeding and thereafter by the return of spontaneous menses and normoprolactinemia. An empty sella was subsequently reported in $^3/_9$ women with intrasellar prolactinomas who completed at least 2 bromocriptine-facilitated pregnancies (Daya et al.[25]). In a recent review of the natural history of microprolactinomas, Crosigniani et al.[24] reported 18% "cure" rate of microprolactinomas in patients with bromocriptine- (or other dopamine agonists) facilitated pregnancy pointing out the favourable influence of bromocriptine facilitated pregnancy on the natural history of prolactinomas.

Medical Treatment of Macroprolactinomas: Effects of Bromocriptine

Important evidence of the therapeutic efficacy of bromocriptine administered prior to surgery, after pituitary surgery and as primary therapy has accumulated.

In patients in whom *bromocriptine was administered* for up to 6 weeks *prior to transsphenoidal surgery*[4, 32], marked suppression of the PRL secretion and tumour shrinkage occurred and the tumour appeared to be softer and more fluid facilitating removal by suction[4, 32]. The duration of bromocriptine therapy prior to surgery was considered very important for the outcome of the adenomectomy. Ideally, surgery should be performed when maximal shrinkage of the tumour occurred. To extend the treatment beyond 6 months might make surgical excision difficult as a result of fibrosis which may occur and surgical cure of the tumour impossible to achieve[4, 32].

The efficacy of the *long-term treatment with bromocriptine* or other dopamine agonists in patients in whom hyperprolactinemia persists *postoperatively* is also well documented. Thus treatment with bromocriptine following surgery usually achieves a normalization of PRL levels and a resumption of gonadic function in both men and women, provided that the gonadotrophic cells have not been destroyed by tumour expansion or by surgical manipulation.

The use of bromocriptine as primary and long-term therapy for macroprolactinomas has become a widely accepted alternative to surgery with considerable but variable shrinkage, in some cases to almost complete disappearance of macroprolactinomas[61, 66, 95]. In a series of 18 patients with pituitary macroadenomas and hyperprolactinemia, Wass et al.[95] reported unequivocal tumour shrinkage in 12 patients with partially empty sella in 2 patients, 16 and 22 months after starting treatment with bromocriptine respectively.

In a larger series of 33 patients, 15 treated with bromocriptine, 12 with lisuride and 3 with consecutive courses of either drug for periods of 6 to 64 months, Liuzzi et al.[61] reported tumour shrinkage in 20 patients (60.5%) accompanied by normalization or marked reduction of PRL plasma levels. In $^7/_{20}$ patients tumour shrinkage was seen within the first month of treatment whereas in the other 13 patients the shrinkage become evident later, up to 8 months after continuous drug administration. The effects of the therapy were maintained in all patients up to 5 years follow-up. Ten of 33 patients did not show changes in tumour size whereas PRL plasma levels fell markedly. Resistance

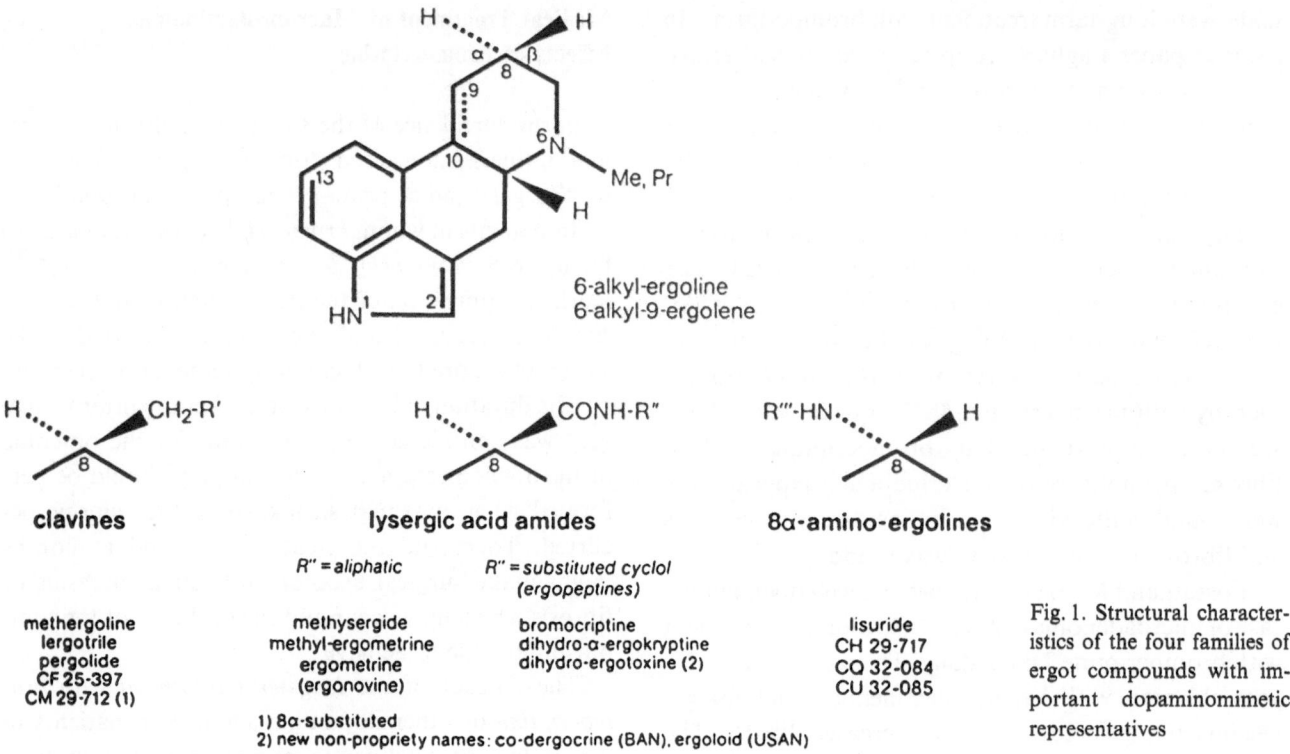

Fig. 1. Structural characteristics of the four families of ergot compounds with important dopaminomimetic representatives

to bromocriptine and/or lisuride was recorded in $^3/_{33}$ patients in the series of Luizzi et al.[61]. More recently Molitch et al.[66] published the results of a multicentre prospective study carried out in 27 patients. Normal PRL levels were reached in $^{18}/_{27}$ patients and various degrees of reduction in the tumour size was recorded in all patients. In 13 patients (46%) tumour size was reduced by greater than 50%, in 5 patients (18%) by about 50%, and in 9 patients (36%) by approximately 10–25%[66]. The time to reach maximal size reduction was 6 weeks in 6 patients, 6 months in 13 patients and 12–15 months in 8 patients[66].

Inhibitors of PRL Secretion Used in the Treatment of Prolactinomas

Drugs which suppress PRL secretion by a dopaminergic mechanism are found in several chemical classes. Of these, the ergolene and ergoline compounds, ergot alkaloids and related compounds (see Fig. 1) have been the most successful classes[36]. Among the dopaminergic drugs, in which the structure does not fit in with the conservative ergot alkaloids, is the octahydrobenz (g) quinoline compound, CV 205-502, selected recently for clinical use (Fig. 2)[37]. Lergotrile, mesulergin (CU 32-085), methergoline, lisuride, pergolide, CQP 201-403, CV 205-502 and cabergoline have been used extensively in the treatment of hyperprolactinemic states. All are orally effective and could be administered b.i.d. or t.i.d. (bromocriptine, lisuride, CU 32-085), once a day (bromocriptine, pergolide, CQP 201-403, CV 205-502) or once a week (cabergoline). Lisuride (0.2 mg t.i.d.) produced a slightly more rapid decrease of PRL secretion than bromocriptine but its effect was less prolonged after a single dose compared to bromocriptine. Neither the therapeutic nor the safety profile of lisuride showed any superiority over bromocriptine in the treatment of hyperprolactinemic states and prolactinomas[13, 61].

Pergolide mesylate, the first compound with long-lasting activity used clinically, is active orally in a dose of 25 to 100 µg/day. Pergolide, like lisuride and bromocriptine, acts directly at the level of the pituitary and PRL-secreting tumours to inhibit PRL secretion. As the therapeutic effects and adverse events are similar to those demonstrated with bromocriptine[11, 48] the key advantage of this drug appears to be its extended half-life and once a day administration.

Among the new molecules which are presently under clinical investigation the most promising results have been reported with: *CQP 201-403*, (8-(N,N-Diethyl-sulfamoyl amino)-6-4-propyleroline-hydrochloride), clinically used at single doses of 0.01–0.03 mg/day[80]; *CV 205-502*, (3-N,N-diethylsulfamyl amino-1,2,3,4,4 a,5,10,10 aβ-octahydro-6-hydroxy-l-propyl

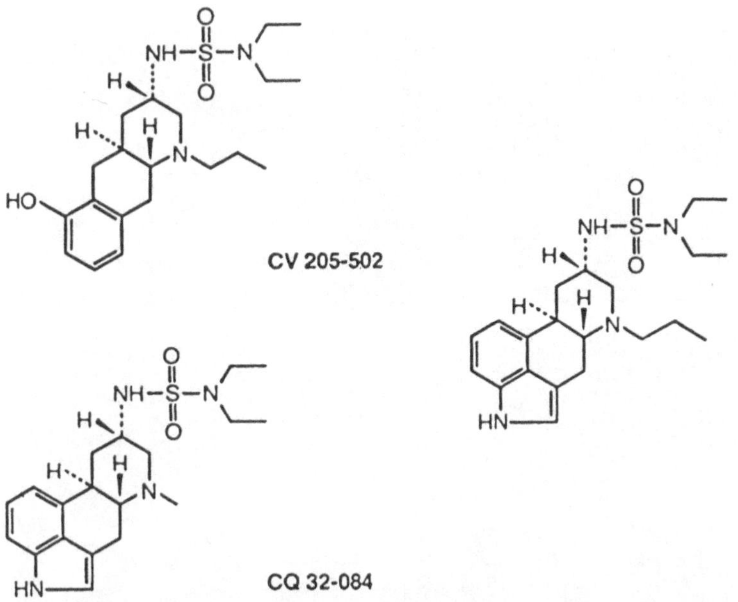

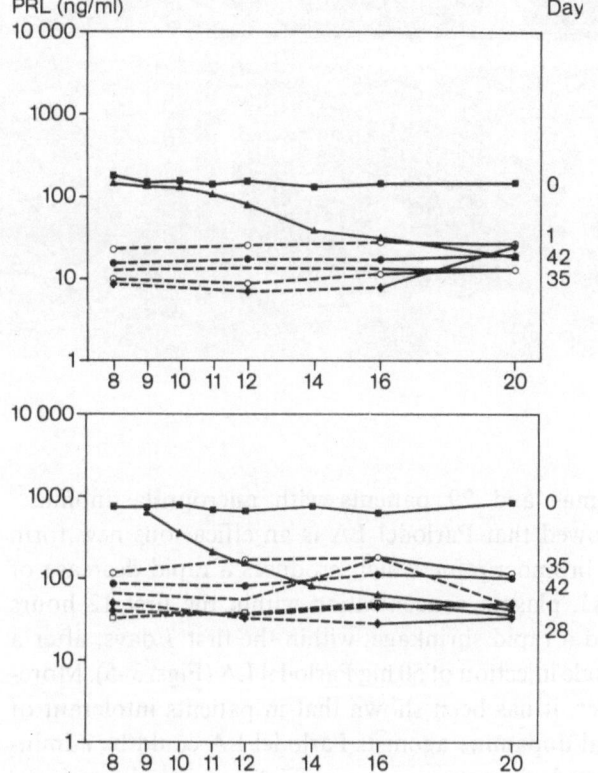

Fig. 2. Chemical structures of CQ 32-084 and its derivatives CQP 201-403 and the non-ergoline derivative CV 205-502

benzoquinoline hydrochloride) used at single daily doses of 0.05–0.1 mg[77] and *Cabergoline* (1(6-allylergolin-8p-yl)carbonyl)-1-(3-dimethylamino(propyl)-3-ethylurea used weekly at doses of 0.3–0.6 mg[34, 65].

Data available to date suggest a good tolerability for all these new compounds, good efficacy in suppressing PRL secretion and good acceptance by patients. The evidence of tumour shrinkage is scarce for all these new compounds which are still in early clinical development.

New Galenical Forms of Bromocriptine in the Treatment of Prolactinomas

Treatment with oral bromocriptine and other dopamine agonists suppresses PRL secretion and shrinks PRL-secreting tumours and made possible the medical management of prolactinomas. However the initial adverse effects encountered, particularly if full doses are used, at the start of treatment and the limited compliance seen in some patients during long-term treatment led to the search for new non-invasive approaches.

Recently, two injectable forms of bromocriptine (Parlodel Long-Acting (LA) and Long-Acting Repeatable (LAR), Sandoz) have been developed for intramuscular injections.

Parlodel LA contains bromocriptine microspheres prepared according to a spray-drying technique which employs polylactic acid as wall material. The bromocriptine-containing microspheres are injected deep intragluteally where they slowly release their content over 6–8 weeks. Therapeutic plasma concentrations of bromocriptine (1–1.5 ng/ml) are reached within the first

hours after injection of Parlodel LA, as was shown bay studies carried out in normal volunteers[54]. Marked reduction of PRL secretion was noted 3 hours after injection and a sustained suppression of PRL secretion was found for about six weeks after a single i. m. injection of 50 mg Parlodel LA in normal volunteers[54]. Results recorded in 87 patients with macroprolacti-

Fig. 3. PRL pattern in two patients prior to (day 1) and up to day 42 after the injection of bromocriptine LA administered at 8 a.m. on day 1

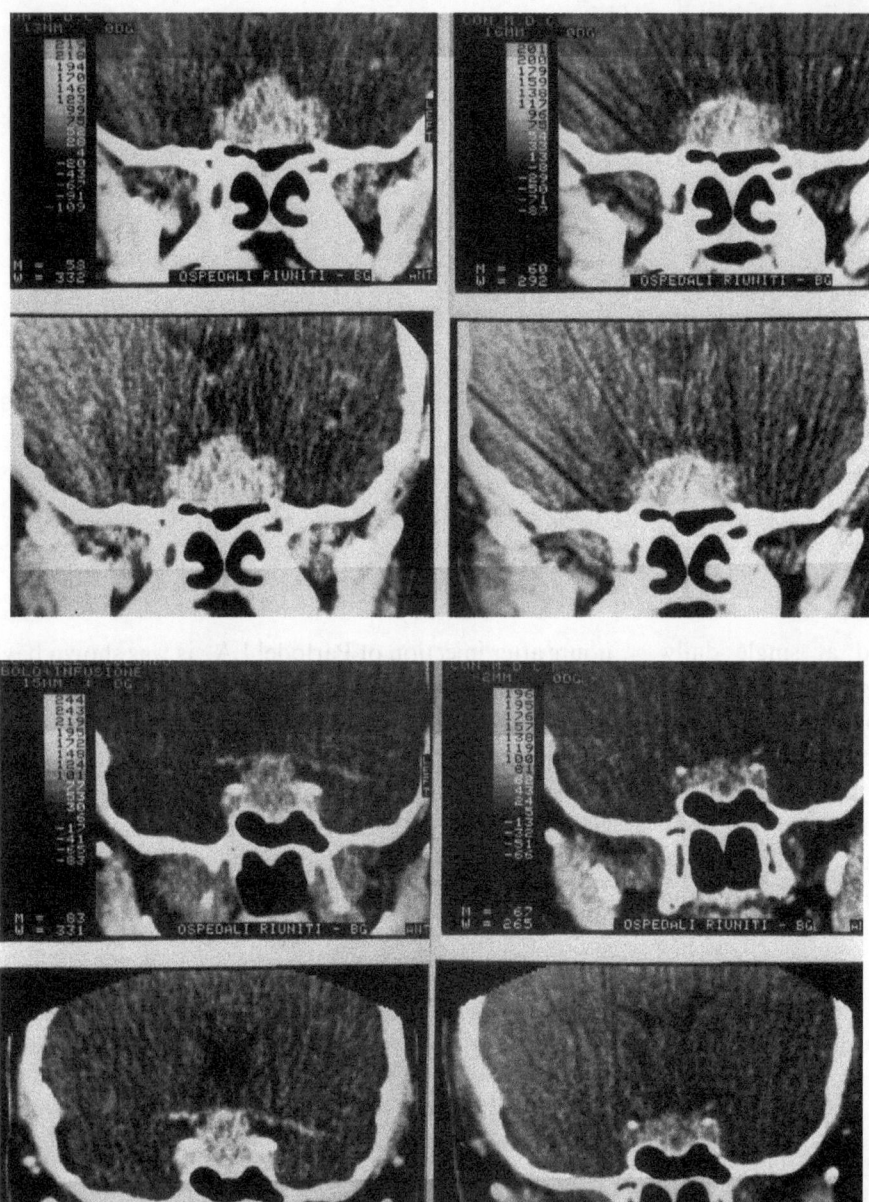

Fig. 4. Tumour shrinkage documented by coronal CT before treatment (left panel) and 15 days after bromocriptine LA (right panel) in one patient with supra- and parasellar extension of the prolactinoma

Fig. 5. Marked tumour shrinkage documented by coronal CT before treatment (left panel) and 21 days after bromocriptine LA (right panel) in one patient with suprasellar extension of prolactinoma

nomas and 29 patients with microprolactinomas[55] showed that Parlodel LA is an efficacious new form of bromocriptine which produces a rapid decrease of PRL plasma concentration within the first 12 hours and a rapid shrinkage, within the first 7 days, after a single injection of 50 mg Parlodel LA (Figs. 3–5). Moreover, it has been shown that in patients intolerant of oral dopamine agonists Parlodel LA could be administered without any adverse effects or with minor, short-lasting adverse effects occurring only within the first hours after the injection[55]. Transfer to oral bromo-

criptine was possible with or without minor adverse effects in most patients[44, 55]. Parlodel LA, the first injectable form of bromocriptine containing polylactic acid that metabolizes very slowly (half life in animal models: approx. 6 months) and hence was introduced on the market only for single injections as initial treatment of prolactinomas. The need for a new injectable form for repeated administrations is clear and a new long-acting injectable form of Parlodel, Parlodel LAR (Long-Acting Repeatable), was developed for monthly injections. Parlodel LAR contains bromocriptine-microspheres which employ D,L-polylactide-coglycolide glucose as carrier material. The in vitro and in vivo experiments showed that the degradation of the polymer was nearly complete after 52 days.

Data are available in 26 patients (16 men, 10 women, aged 19–65 years) with macroprolactinomas and 20 patients (all but one woman, aged 18–45 years) with microprolactinomas treated for 2–12 months with Parlodel LAR. All patients received a 50 mg dose initially and 50, 75 or 100 mg/month thereafter for obtaining maximal inhibition/normalization of PRL plasma levels throughout a 4-week period after each injection[56].

Efficacy of Parlodel LAR in Patients with Macroprolactinomas

PRL plasma levels, which ranged from 207 to 40,340 ng/ml, were suppressed by 50–70% from pretreatment levels in all patients within 12 hours after the first 50 mg dose of Parlodel LAR. In $^{12}/_{26}$ patients PRL plasma levels fell markedly to within the normal range (< 25 mg/ml) after the first injection of Parlodel LAR and remained suppressed until day 28 as illustrated in one representative case in Fig. 6. All patients

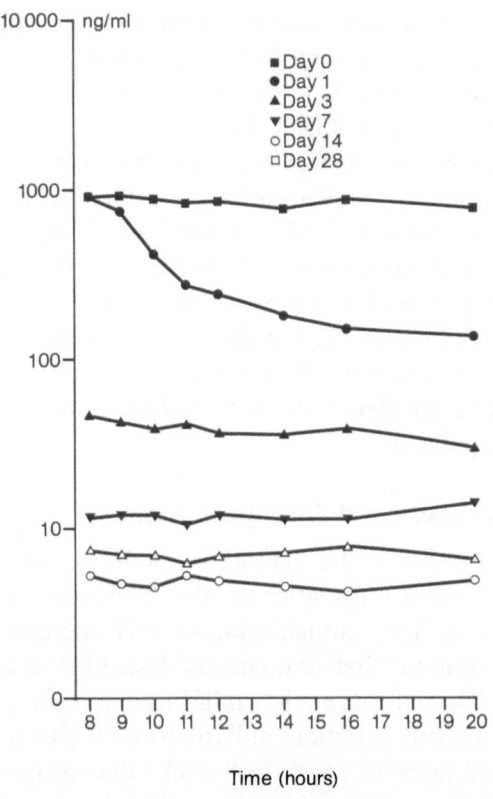

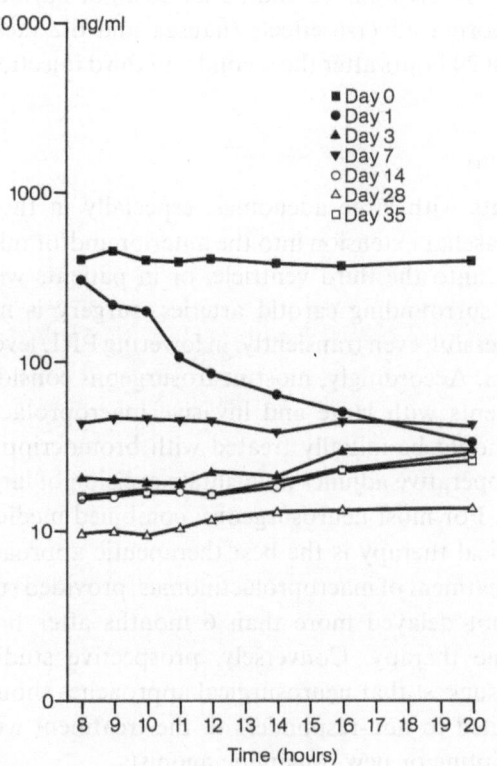

Fig. 6. **PRL** plasma levels measured eight times daily from 8 a.m. to 8 p.m. to assess the PRL profile before (day 0) and on days 1, 3, 7, 14, 28 and 35 after the first 50 mg dose of Parlodel LAR in one patient with macroprolactinoma

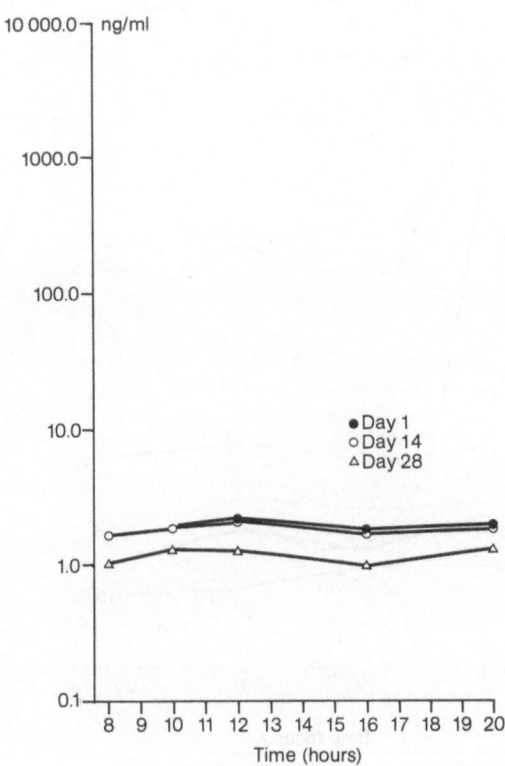

Fig. 7. **PRL** plasma concentrations (ng/ml) in one patient with macroprolactinoma after the first (50 mg) and fourth (50 mg) injection of Parlodel LAR

but 3, showed a progressive and sustained suppression of PRL plasma levels during the 2–12-month treatment period to within normal levels, as illustrated in one representative case in Fig. 7. Shrinkage of PRL-secreting macroadenomas was documented by CT-scan examinations in $^{20}/_{26}$ patients on day 5–7 after the first 50 mg dose of Parlodel LAR; the reduction in size of pituitary tumour ranged from 20% to greater than 50% of the original size. Further reduction in tumour size was recorded by day 28 after the first injection in 8 patients, as demonstrated by CT scan after 6 months treatment in 12 of 22 patients and after 12-month treatment in $^{4}/_{6}$ patients.

Efficacy in Patients with Microprolactinomas

In all but 2 patients PRL plasma levels decreased to within the normal range after the first 50 mg dose of Parlodel LAR. The normalization of PRL secretion occurred within the first 12 hours and lasted for 28 to 72 days in $^{18}/_{20}$ patients as shown in Fig. 8 in one representative patient. Injections of Parlodel LAR (50 mg) administered thereafter at 4- to 8-week intervals provided a continuous suppression to within the normal range for the whole follow-up period. In $^{2}/_{20}$ patients with microprolactinomas the dose of Parlodel LAR had to be increased to 100 mg/month for a long-lasting normalization (4 weeks) of PRL plasma concentrations. In all patients galactorrhea disappeared and menses resumed within 2 months after starting the treatment with Parlodel LAR. The injection caused minimal and short-lasting local reactions. Pain (slight or moderate) at the injection site was reported by $^{10}/_{46}$ patients, soreness by $^{5}/_{46}$ patients and rash by $^{6}/_{46}$ patients after the first injection. Repeated administrations of Parlodel LAR were locally well tolerated. Slight pain, rash or soreness being recorded only in $^{6}/_{46}$ patients.

Adverse effects were reported after the first injection of Parlodel LAR in $^{16}/_{46}$ patients with prolactinomas ($^{9}/_{26}$ with macroprolactinomas and $^{7}/_{20}$ with micrprolactinomas) and consisted of nausea ($^{15}/_{46}$) and headache ($^{8}/_{46}$) followed by vomiting ($^{5}/_{46}$), dizziness ($^{5}/_{46}$), sleepiness ($^{2}/_{46}$) and nasal congestion ($^{1}/_{46}$). These adverse effects were transient, lasting less than 24 hours in $^{13}/_{16}$ patients. One patient reported moderate dizziness for 14 days, an effect which disappeared without any treatment. One patient reported headache for 10 days and another patient asthenia for 2 days. Only $^{4}/_{46}$ patients receiving 2–12 additional doses of Parlodel LAR reported adverse effects (nausea and dizziness) in the first 24 hours after the second and third injection.

Conclusions

In patients with large adenomas, especially in those with extrasellar extension into the anterior and/or middle fossa, into the third ventricle, or in patients with tumours surrounding carotid arteries, surgery is not very successful, even transiently, in lowering PRL levels to normal. Accordingly, most neurosurgeons consider that patients with large and invasive macroprolactinomas should be initially treated with bromocriptine as a pre-operative adjunct to facilitate excision of large tumours. For most neurosurgeons, combined medical and surgical therapy is the best therapeutic approach for the treatment of macroprolactinomas, provided surgery is not delayed more than 6 months after bromocriptine therapy. Conversely, prospective studies strongly suggest that neurosurgical approaches should be restricted to non-responders to the treatment with bromocriptine or new dopamine agonists.

The place of medical therapy in the management of microprolactinomas has become more important since evidence was collected showing recurence of hyperprolactinemia in patients initially thought to be cured by

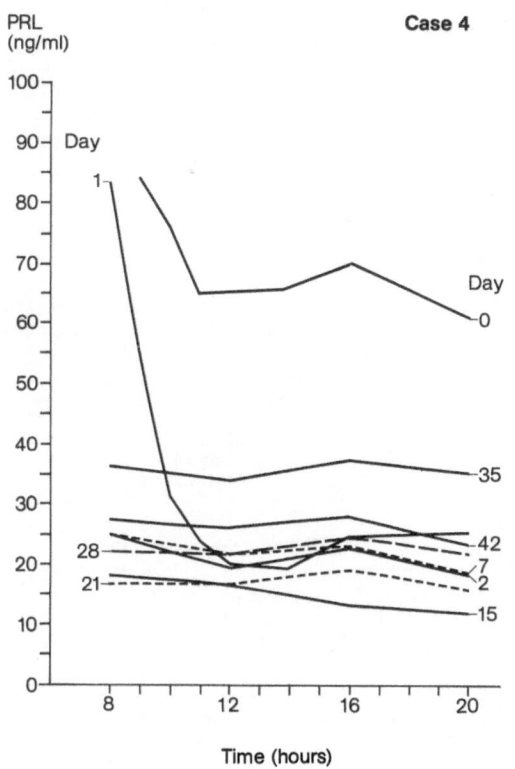

Fig. 8. PRL plasma levels measured eight times daily from 8 a.m. to 8 p.m. to assess the PRL profile before (day 0) and on days 1, 3, 7, 14 and 28 after the first 50 mg dose of Parlodel LAR in one patient with microprolactinoma

surgery. The medical treatment is a long-term treatment which restores to normal the gonadal function only during the treatment period. Evidence is too limited as yet to define the cure rate of such microprolactinomas by medical therapy.

In the management of prolactinomas progress was made recently by the introduction of the injectable, long-acting forms of bromocriptine, Parlodel LA and Parlodel LAR. Based on the results obtained with Parlodel LAR in patients with prolactinomas, a new approach in the treatment of prolactinomas is proposed, namely, to initiate treatment of patients with PRL-secreting adenomas with this long-acting bromocriptine (Parlodel LAR) in order to obtain a rapid decrease of PRL plasma levels and a rapid shrinkage of the tumour and to continue the chronic treatment with monthly injections of 50 to 100 mg Parlodel LAR. It is possible to establish early whether the patients will be responsive or not to a dopamine agonist.

Medical Approaches to the Management of GH-Secreting Tumours

The introduction of dopamine agonists in the management of acromegaly, a disease produced by GH-secreting pituitary tumours, more than 10 years ago opened the non-invasive approach to this severe endocrine disease. More recently the development of a long-acting somatostatin analogue (Sandostatin, SMS 201-995, Sandoz) and early reports of its clinical efficacy added a new and very important therapeutic tool to the medical management of acromegaly.

The use of the somatostatin analog is based on important knowledge accumulated on the control of GH secretion in humans. Indeed in man, as in all mammals, somatostatin, the natural hypothalamic neurohormone, is the physiological inhibitor of GH secretion stimulated by all known stimuli[35, 78]. Like somatostatin, SMS 201-995 inhibits GH secretion stimulated by arginine, exercise, sleep and insulin-induced hypoglycemia[76]. Whereas the clinical rationale for somatostatin analog treatment in acromegalics is obvious, the use of dopamine agonists is based on the paradoxical decrease in growth hormone (GH) secretion after the administration of L-DOPA in patients with acromegaly reported by Liuzzi et al. in 1972[58].

The medical treatment of acromegaly takes second place to transphenoidal-adenomectomy with or without pituitary radiation except where such therapy is unsuccessful. The outcome of neurosurgery depends on the size of the tumour: patients with tumour of less than 10 mm in diameter (microadenomas) have approximately an 80% chance of cure, whereas in macroadenomas, particularly when invasive, the cure rate may be as low as 52%. Surgery may be followed by secondary hypopituitarism and recurrence of the disease.

Conventional external irradiation is an effective treatment in patients with acromegaly with moderately increased GH plasma concentrations. The normalization of GH levels could occur in about 30–80% of the patients within 6–24 months, in 20% of acromegalics being delayed as long as 5–10 years. Pituitary implantation of yttrium or proton beam therapy causes a good clinical and biological effect in 27–50% of patients but the risk of secondary hypopituitarism is very high.

When considering the place of medical treatment with (1) dopamine agonists and (2) somatostatin (Sandostatin) the following ideal objectives will be considered: a) correction of the hormonal (GH) excess and its clinical and biological consequences, b) reduction of the tumour mass and c) prevention of impairment of the remaining pituitary tissue hormone secretion.

Dopamine Agonists in the Treatment of GH-secreting Tumours

Most clinical experience has been obtained with bromocriptine. The therapeutic effects of lisuride, pergolide, mesulergine and other dopamine agonists have been proven to be equally good, thus in this short review we shall refer only to bromocriptine data. Single doses of 2.5 mg bromocriptine significantly decrease plasma GH levels in most acromegalic patients[59, 84] and the stable and significant reduction of GH levels during long-term treatment with bromocriptine is amply documented in acromegalic patients[6, 9, 15, 18, 60, 69, 70]. Bromocriptine can reduce GH hypersecretion and control its metabolic consequences in at least 50% of acromegalics. However when given alone it restores GH levels to normal in relatively few patients. Wass et al.[93] reported a GH concentration of less than 10 mU/l in only 15 of 73 patients (20%) treated with bromocriptine despite objective improvement in most patients. Lamberts et al.[51] claimed that patients who respond well to bromocriptine therapy are more likely to hypersecrete both GH and PRL whereas other groups reported good clinical and biochemical effects in patients with isolated GH-secreting tumours[6, 9, 15, 18, 60, 69, 70].

The few reports of improvement in visual field defects in acromegaly patients under bromocriptine treatment[64, 94, 99] prompted several groups to study tumour shrinkage in acromegalics. McGregor[64] failed to observe any change in tumour size in four acromegalic patients studied by CT before and after 3 months of therapy with bromocriptine (20 mg/day) whereas Wollesen et al.[99], Liuzzi et al.[61] and Wass et al.[94] reported reduction of the tumour size in 13% to 50% of acromegalic patients. The fact that reduction in the size of GH-secreting adenomas was recorded only in acromegalic patients responsive in terms of GH secretion is consistent with the view that the presence of dopamine D_2 receptors on the GH-secreting tumours is a prerequisite for obtaining tumour shrinkage by bromocriptine. The presence of dopamine D_2 receptors on GH-secreting pituitary adenomas obtained from patients who were pre-operatively responsive to either DA infusion or DA agonists has been documented by several groups[71, 83]. However, in a recent in vitro study, Brassion et al.[14] showed qualitative differences between the dopamine receptors of the tumoural PRL- and GH-secreting cells in humans, the receptors of somatotrophs exhibiting less affinity for apomorphine, DA and spiperone than those of the lactotrophs. The lack of DA receptors and/or their reduced affinity for DA could explain the lack of therapeutic effects reported for bromocriptine in some acromegalics.

New Galenical Forms of Bromocriptine in the Treatment of Acromegaly

The two injectable long-acting forms of bromocriptine, Parlodel LA and Parlodel LAR, used in the treatment of patients with prolactinomas, have been assessed in patients with acromegaly. Twenty-four patients, 10 men and 14 women with acromegaly were treated with Parlodel LA. Eight of these patients were hyperprolactinemic in addition to having symptoms and signs of acromegaly. Five women (36%) and 1 man (10%) presented with galactorrhea at the time of baseline evaluation. Thirteen women (93%) presented with menstrual disturbances such as amenorrhea (10 patients) and oligomenorrhea (3 patients). Nine patients, 3 women and 6 men reported an absence or decreased libido and 5 men had below normal level of potency. Symptoms/signs of active acromegaly were recorded in all patients. All patients had pituitary tumours documented by CT scan, $^{12}/_{24}$ patients had a suprasellar extension of the tumour and 3 of the patients with suprasellar extension of the tumour had a definite impairment of the visual fields. Visual field disturbances were detected in a further 4 patients who had no evident suprasellar extension of the tumour. Other endocrine abnormalities due to pituitary insufficiency (tumour

mass effect) were observed in $^7/_{24}$ patients (29%). Hypertension, a frequent cardiovascular complication of acromegaly, was present in $^{10}/_{24}$ (42%) patient prior to the administration of Parlodel LA. One patient had angina pectoris but none of the patients had a past history of myocardial infarction.

GH fell markedly in $^{16}/_{24}$ patients (67%) and reached normal values (equal to or less than 5 ng/ml) in $^6/_{24}$ patients (25%). In $^4/_{24}$ patients (17%) GH secretion was suppressed throughout the day for 28–42 days after a single injection of 50 mg Parlodel LA. While the proportion of patients with long-lasting (4–6 weeks) normalization of GH secretion is relatively low ($^4/_{24}$) this long-lasting result obtained with Parlodel LA compares well with the best therapeutic responses reported for bromocriptine treatment in acromegaly.

PRL secretion was normalized in $^7/_8$ acromegalic patients within 24 hours after the injection of Parlodel LA and the suppression lasted for 42 days in all 7 patients. In one patient PRL secretion was only partially suppressed from 3,930 ng/ml to nadir concentrations of 101 ng/ml by day 3. The suppression of PRL secretion was long-lasting and by day 42, PRL plasma concentration was 523 ng/ml (13% of the initial levels). This patient had a very large macroadenoma with marked visual field defects and bilateral optic atrophy produced by the chronic chiasmal compression. This patient, who presented with relapse of acromegalic symptoms and an enlargement of the tumour after irradiation, was also known to be a poor responder to oral Parlodel administered up to 22.5 mg daily. After the administration of Parlodel LA the patient had, in addition to the partial suppression of PRL secretion, a normalization of GH plasma levels lasting for 42 days and tumour shrinkage (30–50% of the initial size) detected by day 42. The therapeutic effects obtained with Parlodel LA were maintained with the subsequent oral treatment with Parlodel (7.5 mg/day).

Decrease or disappearance of symptoms/signs of acromegaly was recorded in most patients as shown below:

Decrease/disappearance of hyperhidrosis	$^{14}/_{20}$ patients
Reduction of palmar soft-tissue thickening	$^{11}/_{22}$ patients
Disappearance of headache	$^{10}/_{12}$ patients
Decrease of muscular weakness	$^9/_{14}$ patients
Decrease of arthritic complaints	$^7/_{15}$ patients
Disappearance of menstrual abnormalities	$^2/_7$ patients

Disappearance of galactorrhea $^5/_6$ patients
Improved libido $^3/_9$ patients
Improved potency $^3/_5$ patients
Improvement of carpal tunnel syn- $^3/_5$ patients
drome
Improvement of visual fields $^5/_7$ patients
Normalization of the blood pressure $^5/_{19}$ patients

Shrinkage of the tumour was documented by CT scan in $^4/_{22}$ patients within 42 days after the injection of Parlodel LA. Three of these 4 patients with documented shrinkage of the tumour (30–50% of the initial size) had acromegaly and hyperprolactinemia. Six acromegalic patients received a second injection of Parlodel LA. Long-lasting suppression of GH secretion was seen in $^4/_6$ patients, GH secretion being suppressed to within the normal range for 28 days in one patient.

Thus, results obtained with a single injection of Parlodel LA compared favourably with those obtained with Parlodel administered orally. The rapid shrinkage of the GH secreting-pituitary tumours in some patients and the long-lasting (4–6 weeks) normalization of GH secretion in patients who respond well to the treatment with dopamine agonists encouraged the assessment of Parlodel LAR (50 mg or 100 mg doses) administered once a month in patients with acromegaly. Preliminary results of on-going studies suggest that one could replace the oral treatment with monthly injections of Parlodel LAR in patients who show a clinical and biochemical improvement on long-term treatment with oral bromocriptine.

The Somatostatin Analogue, Sandostatin (SMS 201-995) in the Treatment of GH-secreting Tumours

Plewe et al.[74]; Lamberts et al.[52]; Chiodini et al.[19]; Cozzi et al.[23]; Ch'ng et al.[21] and Plewe et al.[75] were the first to report on the therapeutic effects of the somatostatin analog SMS 201-995. Chronic therapy results in a rapid and long-lasting suppression of the GH secretion and the metabolic consequences of the GH hypersecretion. Daily dosages of 100–1,500 μg given in 2–3 subcutaneous injections have been reported to reduce GH and Somatomedin C (SmC) levels in the majority of the patients[19, 21, 52, 74]. The supression of GH secretion lasts as long as SMS 201-995 is administered.

In a recent study Lamberts et al.[53] report results in 10 acromegalic patients treated for 4–27 months with SMS 205-995. A marked clinical improvement was noted during the first weeks of therapy, clinical improvement which was more pronounced that the fall in serum GH and SmC levels. Mean GH plasma levels (mean of 19 samples collected in a 24 h period) showed concentrations below 5 μg/l in $^6/_{10}$ patients whereas serum SmC levels decreased to within the normal range in $^5/_{10}$ patients. Three of the six patients in whom CT scan evaluations were made showed a slight decrease of the maximal diameter of the adenoma after 14–16 weeks therapy with SMS 205-995. This long-term study confirmed the observation made by Pieters et al.[73] that the acute response of serum GH levels to 50 μg SMS 205-995 could predict the response to long-term treatment. No desensitization to the effect of SMS 201-995 on GH secretion was reported after long-term treatment and no important adverse effects were noted. Transient steatorrhea and abdominal discomfort occurred in $^2/_{10}$ patients on long-term treatment[53].

In another recent paper Chiodini et al.[20] report results obtained with SMS 205-995, 100–1,500 μg/day, administered for 3–27 months. A rapid decrease in GH and Sm-C plasma levels was noted by the second day of treatment with nadir levels noted after 15 days of treatment. Thereafter there were no further significant changes of GH and SmC levels even in patients treated for up to 27 months. In $^{22}/_{26}$ patients GH and SmC plasma levels decreased to more than 50% of the initial levels whereas resistance to SMS 205-995 was recorded in $^4/_{26}$ patients. In one of these 4 patients no decrease in GH levels was seen even at high doses of SMS 205-995 e.g. 1,500 μg/day. The comparison between the long-term effects of SMS 205-995 (100–300 μg/day) and bromocriptine performed by Chiodini et al. in 13 patients[20] showed that SMS 205-995 was more efficacious in lowering GH and SmC plasma levels in $^{11}/_{13}$ patients. The incidence of poor responders, $^2/_{13}$ to SMS 205-995 and $^6/_{13}$ patients to oral bromocriptine, pointed to the better therapeutic effects of SMS 205-995 in acromegalic patients.

A comparison between long-term treatment with multiple daily injections and continuous infusion of the same dose of SMS 205-995 (100–600 μg daily) was also carried out by Chiodini et al.[20]. A significant decrease of GH and SmC plasma levels was noted with either kind of treatment within the second day of treatment but, during continuous infusion of SMS 205-930, a more marked decrease of GH and SmC plasma levels was seen. Moreover, on long-term treatment with continuous infusion of SMS 205-995 all patients showed a consistent decrease of GH and SmC levels to within the normal range, an effect which lasted during the whole period of treatment. The suppression of GH secretion was less stable and consistent during the multiple daily injections with SMS 205-995. The discontinuation of each type of treatment with SMS 205-995 was followed by a rapid return (24–36 hours) of GH and SmC to the pre-treatment levels[20].

Clinically and metabolically all patients improved on both types of drug administration. Diabetes improved and, in two of the three diabetics, insulin therapy could be stopped[20]. In the paper by Chiodini et al.[20] reduction in tumour size was recorded by CT scan examinations in 10/20 patients with large tumours, the incidence of shrinkage being similar to that report by other groups[3, 52, 53].

The results reported to date with SMS 205-995 clearly show that this long-acting somatostatin analog is the best available medical treatment for acromegalic patients. Administered as multiple doses (110–500 μg t.i.d. sc) or as continuous infusion of the same dose, SMS 205-995 suppresses GH secretion in 85% of acromegalic patients. A marked fall in SmC concentrations and improvement of the clinical symptoms and metabolic alterations which characterize acromegaly were reported even in patients who did not show a suppression of GH secretion to within the normal range. No desensitization of the pituitary receptors for somatostatin was reported during long-term treatment with SMS 205-995.

The fact that after discontinuation of the SMS 205-995 therapy, GH plasma concentrations rise to pretreatment levels within 24–36 hours suggest that the mechanism of action of the drug is not identical to that of bromocriptine in prolactinomas. Indeed, George et al.[42] reported that pretreatment of one acromegalic patient for 10 days with SMS 201-995 resulted in a decrease of tumour cell size, nuclear and cytoplasmic areas and an increased volume density of lysosomes, possibly indicating increased crinophagy (intracellular break-down) of GH. Therefore, a slight decrease in tumour size during therapy might well reflect only a decrease in the size of the individual tumour cells (caused by lower activity and hormone content), rather than a cytotoxic or vascular effect of the drug. This concept is also supported by the rapid return to pretreatment circulating GH levels after the discontinuation of SMS 201-995 treatment.

Combined Treatment with SMS 205-995 and Bromocriptine in Acromegaly

In acromegalic patients with only partial suppression of GH and SmC plasma concentrations on long-term treatment with SMS 205-995 the addition of oral bromocriptine was reported to improve or normalize GH

and SmC concentrations[19, 20]. More recently, ongoing studies (Montini et al., unpublished data) showed a suppression of GH secretion to within the normal range by combining Parlodel LAR (100 mg once a month i.m.) to SMS 205-995 treatment in 2 acromegalics who showed only a partial improvement on long-term therapy with SMS 205-995.

Conclusion

The introduction of dopamine agonists and of a long-acting somatostatin analog in the treatment of PRL-and/or GH-secreting pituitary tumours has added a new dimension to the therapy of tumoural hyperprolactinemia and acromegaly and provides a medical treatment for patients with pituitary tumours who cannot be cured by surgery.

References

1. Aubourgh MD, Derome PJ, Peillon F, Jedynak CP, Visot A, Le Gentil P, Balagura S, Guiot G (1980) Endocrine outcome after transsphenoidal adenomectomy for prolactinoma: prolactin levels and tumour size as predicting factors. Surg Neurol 14: 141–152
2. Barbarino A, De Marinis L, Anile C (1982) Dopaminergic mechanism regulating prolactin secretion in patients with prolactin-secreting pituitary adenoma. Long-term studies after selective transsphenoidal surgery. Metabolism 31: 1100–1104
3. Barnard LB (1986) Reduction of pituitary tumour size in acromegaly treated with a somatostatin analogue (SMS 201-995): Abst 983 in The 68th Annual Meeting of the Endocrine Society, Anaheim
4. Barrow DL, Tindall GT, Kovacs K, Thorner MO, Horvath E (1984) Clinical and pathological effects of bromocriptine on prolactin secreting and other pituitary tumours. J Neurosurg 60: 1–7
5. Bassetti M, Spada A, Pezzo G, Giannattasio G (1984) Bromocriptine treatment reduces the cell size in human macroprolactinomas: A morphometric study. J Clin Endocr 58: 268–273
6. Belforte L, Camanni F, Chiodini PG, Liuzzi A, Massara F, Molinatti GM, Muller EE, Silvestrini F (1977) Long-term treatment with 2 Br-alpha ergocryptine in acromegaly. Acta Endocrinol 85: 235–241
7. Bergh T, Nillius SJ (1982) Prolactinomas: follow-up of medical treatment: In: Molinatti GM (ed) Clinical problem: Microprolactinoma. Excerpta Medica, Amsterdam, pp 115–130
8. Besser GM, Parke L, Edwards CRW, Forsyth IA, McNeilly AS (1972) Galactorrhoea: successful treatment with reduction of plasma prolactin levels by bromergocryptine. Br Med J 3: 669–672
9. Besser GM, Wass JAH (1984) The medical treatment of acromegaly. In: Black PL, Zervas NT, Ridgway EC, Martin JB (eds) Secretory tumours of the pituitary gland. Raven Press, New York, pp 155–168
10. Besser GM, Wass JAH, Grossman A, Ross R, Doniach I, Jones AE, Rees LH (1985) Clinical and therapeutic aspects of hyperprolactinemia. In: MacLeod RM, Thorner MO, Scapagnini U (eds) Prolactin. Basic and clinical correlates. Liviana Press, Padova and Springer, Berlin Heidelberg New York, pp 833–847

11. Blackwell RE (1985) Diagnosis and management of prolactinomas. Fertil Steril 43: 5–16

12. Bonneville JF, Poulignot D, Cattin F, Couturier M, Mollet E (1982) Computed tomographic demonstration of the effects of bromocriptine on pituitary microadenomas size. Radiology 143: 451–455

13. Bouloux PMG, Besser GM, Grossman A (1987) Clinical evaluation of lysuride in the management of hyperprolactinemia. Br Med J 294: 1323–1324

14. Brassion D, Brandi AM, Nousbaum A, Peillon F (1981) Evidence of dopaminergic receptors in human growth hormone (GH)-secreting pituitary adenomas. Acta Endocrinol Suppl. 243 (Copenh) 97: Abst. 411

15. Camanni F, Massara F, Belforte L, Rosatello A, Molinatti GM (1977) Effect of dopamine on plasma growth hormone and prolactin levels in normal and acromegalic subjects. J Clin Endocrinol Metab 44: 465

16. Camanni F, Ghigo E, Ciccarelli E (1982) Follow-up of 69 patients after pituitary tumour removal for hyperprolactinemia. Excerpta Med Int Congr Ser 584: 205–213

17. Camanni F, Ghigo E, Ciccarelli E, Massara F, Capagnoli C, Molinatti G, Müller EE (1983) Defective regulation of prolactin secretion after successful removal of prolactinomas. J Clin Endocrinol Metab 57: 1270–1279

18. Chiodini PG, Liuzzi A, Botalla L, Oppizzi G, Muller EE, Silvestrini F (1975) Stable reduction of plasma growth hormone (GH) levels during chronic administration of 2-Br-alpha ergocryptine (CB 154) in acromegalic patients. J Clin Endocrinol Metab 40: 705–708

19. Chiodini PG, Cozzi R, Dallabonzana D, Oppizzi G, Verde G, Petroncini M, Liuzzi A, Del Pozo E (1987) Medical treatment of acromegaly with SMS 205-295, a somatostatin analog: a comparison with bromocriptine. J Clin Endocrinol Metab 64: 447–450

20. Chiodini PG, Cozzi R, Dallabonzana D, Oppizi G, Verde G, Petroncini MM, Liuzzi A, Boccardi E, Lancranjan I (1989) Medical treatment of acromegaly, dopaminergic agonists and long-acting somatostatin. In: Müller E, Cocchi D, Locatelli V (eds) Advances in growth hormone and growth factors research. Pythagora Press Roma, Milano, Springer, Berlin Heidelberg New York Tokyo, pp 423–436

21. Ch'ng LJC, Sandler LM, Kraenzlin ME, Burin JM, Joplin CF, Bloom SR (1985) Long-term treatment of acromegaly with a long acting analogue of somatostatin. Br Med J 290: 285–297

22. Cowden EA, Thomson JA (1979) Resolution of hyperprolactinemia after bromocriptine-induced pregnancy. Lancet 1 (letter) 613

23. Cozzi R, Chiodini PG, Dallabonzana D, Petroncini M, Verde G, Oppizzi G, Liuzzi A (1986) Effect of SMS 201-995 administered by minipump infusion in acromegaly. International Conference on Somatostatin, Washington, Abst. 64, p 67

24. Crosignani PG, Mattei AM, Galparoli C, Carena M, Spellecchia D (1987) The natural history of idiopatic hyperprolactinemia and microprolactinomas. In: "Prolattina 1987". Abst. 4. Milano 6–7 Nov

25. Daya S, Shewchuck AB, Bryceland N (1984) The effect of multiparity on intrasellar prolactinomas. Am J Obstet Gynecol 148: 512–515

26. Dietemann JL, Portha C, Cattin F, Mollet E, Bonneville JF (1983) CT follow-up of microprolactinomas during bromocriptine-induced pregnancy. Neuroradiology 25: 133–138

27. Domingue JN, Richmon IL, Wislon CB (1980) Results of surgery in 114 patients with prolactin-secreting pituitary adenomas. Am J Obstet Gynecol 137: 102–107

28. Eversmann T, Fahlbusch R, Rjosk HK, von Werder K (1979) Persisting suppression of prolactin secretion after long-term treatment with bromocriptine in patients with prolactinomas. Acta Endocrinol (Copenh) 92: 413

29. Faglia G, Nissim M, Giannattasio G, Moriondo P, Travaglini P, Ambrosi B, Bernasconi V, Giovanelli MA, Vaccari U (1982) Medical therapy of prolactinomas: effects of bromocriptine on serum PRL levels and tumour size and morphology. In: Molinati GM (ed) A clinical problem: Microprolactinoma. Excerpta Medica, Amsterdam, pp 131–136

30. Fahlbusch R, Giovanelli M, Crosignani PG, Faglia G, Rjosk HK, von Werder K (1980) Differentiated therapy of microprolactinomas: significance of transphenoidal adenomectomy. In: Faglia G, Giovanelli MA, MacLeod RM (eds) Pituitary microadenomas. Academic Press, London, pp 443–456

31. Fahlbush R (1980) Surgical failures in prolactinomas. In: Derome PJ, Jedynak CP, Peillon F (eds) Pituitary adenomas, biology, physiopathology and treatment. Asclepios Publishers, France, pp 273–284

32. Fahlbush R, Buchfelder M (1985) Transphenoidal operations for prolactinomas. In: Auer LM, Leb G, Tscherne G, Urdl W, Walter GF (eds) Prolactinomas. De Gruyter, pp 209–224

33. Faria MAJ, Tindall GT (1982) Transsphenoidal microsurgery for prolactin-secreting pituitary adenomas. Results in 100 women with the amenorrhea-galactorrhea syndrome. J Neurosurg 56: 33–43

34. Ferrari C, Barbieri C, Caldara R, Mucci M, Codecasa F, Paracchi A, Romano C, Boghen M, Dubini A (1986) Long-lasting prolactin lowering effect of cabergoline, a new dopamine agonist in hyperprolactinemic patients. J Clin Endocrinol Metab 63: 941–945

35. Fleischer N, Guillemin R (1976) Hypothalamic hypophysiotropic hormones. In: Parsons JA (ed) Hypothalamic hypophysiotropic hormones. Peptide hormones. Macmillan Press Limited, London, pp 317–335

36. Flückiger E, Briner U, Enz A, Markstein R, Vigouret JM (1983) Dopaminergic ergot compounds: an overview. In: Calne DB, Horrowski R, McDonald W, Wuttke W (eds) Lisuride and other dopamine agonists. Raven Press, New York, pp 1–12

37. Flückiger E (1986) Pharmacological profile of dopaminomimetic drugs. In: Molinatti L, Martini L (eds) Endocrinology '85. Elsevier Science Publishers, North Holland, pp 261–271

38. Fossati P, Mazzuca M (1980) Observation of a male patient with prolactinoma treated with long-term bromocriptine therapy alone. In: Derome PJ, Jedynak CP, Peillon F (eds) Pituitary adenomas. Asclepios Publishers, France, pp 305–306

39. Frantz AG, Cogen PH, Chang CH, Holub DA, Housepian EM (1981) Long-term evaluation of the results of transsphenoidal surgery and radiotherapy in patients with prolactinomas. In: Crosignani P, Robin BL (eds) Endocrinology of human infertility: new aspects. Academic Press, London, pp 161–170

40. Franks S, Jacobs HS (1984) Medical treatment of prolactinomas. In: Jacobs HS (ed) Prolactinomas and pregnancy. Proceedings Special Symposium, XIth World Congress on Fertility and Sterility, Dublin. MTP Press Ltd, Lancaster Boston The Hague Dordrecht, pp 39–43

41. Gen M, Uozumi T, Ohta M, Ito A, Kajiwara H, Mori S (1984) Necrotic changes in prolactinomas after long-term administration of bromocriptine. J Clin Endocrinol 59: 463–470

42. George SR, Kovacs K, Asa SL, Horvath E, Cross EG, Burrow GN (1987) Effect of SMS 201-995, a long-acting somatostatin analogue, on the secretion and morphology of a pituitary growth hormone cell adenoma. J Clin Endocrinol 26: 395–402

43. Giovanelli M, Gaini SM, Tomei G (1982) Follow-up review of microprolactinomas operated in 48 female patients. Excerpta Med Int Congr Ser 584: 189–196

44. Grossman A, Ross R, Wass JAH, Besser GM (1986) Depot-bromocriptine treatment for prolactinomas and acromegaly. J Clin Endocrinol 24: 231–238

45. Hardy J, Beauregard H, Robert F (1978) Prolactin-secreting pituitary adenomas: Transsphenoidal microsurgical treatment. In: Robyn C, Harter M (eds) Progress in prolactin: Physiology and pathology. Elsevier, Science Publishers, North Holland, pp 361–370

46. Hardy J (1980) Ten years after the recognition of pituitary microadenomas. In: Faglia G, Giovanelli MA, MacLeod RM (eds) Pituitary microadenomas. Academic Press, London, pp 7–12

47. Jaquet P, Guibout M, Lucioni J, Grisoli F (1978) Hypothalamo pituitary regulation of prolactin in hypersecreting prolactinomas. In: Robyn C, Harter M (eds) Progress in prolactin: Physiology and pathology. Elsevier, Science Publishers, North Holland, pp 371–382

48. Kleinberg DL (1984) Medical treatment of pituitary tumours. Sem Reprod Endocr 2: 63–72

49. Kovacs K, Horvath E (1979) Pathology of pituitary adenomas. Bull Los Angeles Neurol Soc 42: 92–110

50. Kovacs K, Horvath E (1980) Pituitary adenomas associated with hyperprolactinemia: morphological and immuno-cytological aspects. In: Faglia G, Giovanelli MA, MacLeod RM (eds) Pituitary microadenomas. Academic Press, London, pp 123–131

51. Lamberts SWJ (1984) Antimitotic actions of dopaminergic drugs on human pituitary tumours. In: Müller EE, MacLeod RM (eds) Neuroendocrine perspectives, vol 3. Elsevier Science Publishers, Amsterdam, pp 317–343

52. Lamberts SWJ, Witterlinden P, Verschoor L, Van Dongen KJ, Del Pozo E (1985) Longterm treatment of acromegaly with somatostatin analogue SMS 201-995. N Engl J Med 313–321, 1576

53. Lamberts SWJ, Uitterlinden P, Del Pozo E (1987) SMS 201-995 induces a continuous decline in circulating growth hormone and somatomedin C levels during therapy of agromegalic patients for over two years. J Clin Endocrinol Metab 65: 703–710

54. Lancranjan I, Rolland R, L'Hermite M (1986) Inhibition of lactation with depot-bromocriptine. In: Angeli A, Bradlow HL, Dogliotti L (eds) Endocrinology of the breast: Basic and clinical aspects. Annals of the New York Academy of Sciences, vol 464, pp 473–477

55. Lancranjan I (1987) A new approach to imitate the treatment of patients with prolactinomas. In: Genazzani AR, Volpe A, Facchinetti F (eds) Gynecological endocrinology. The Proceedings of the first International Congress of Gynecological Endocrinology. The Parthenon Publishing Group, Casterton Hall, Carnforth, pp 239–252

56. Lancranjan I, Cavagnini F, Felley Ch, Pagani MD, Montini M, Felber JP (1988) Long-term treatment of prolactinomas with a long-acting injectable form of bromocriptine. In: Genazzani A (ed) Brain and reproduction. The Parthenon Publishing Group Casterton Hall, Carnforth, pp 569–578

57. Landolt AM (1980) Biology of pituitary microadenoma. In: Faglia G, Giovanelli MA, MacLeod RM (eds) Pituitary microadenomas. Academic Press, London, pp 107–112

58. Liuzzi A, Chiodini PG, Botalla L, Cremascoli G, Silvestrini F (1972) Inhibitory effect of L-dopa on GH release in acromegalic patients. J Clin Endocrinol Metab 35: 941–943

59. Liuzzi A, Chiodini PG, Botalla L, Cremascoli G, Muller EE, Silvestrini F (1974) Growth hormone (GH)-releasing activity of TRH and GH-lowering effect of dopaminergic drugs in acromegaly: homogeneity of the two respones. J Clin Endocrinol Metab 38: 910–912

60. Liuzzi A, Chiodini PG, Oppizzi G, Botalla L, Verde G, De Stefano G, Colussi G, Graf KJ (1978) Lisuride hydrogen maleate: evidence for a long-lasting dopaminergic activity in humans. J Clin Endocrinol Metab 46: 196–200

61. Liuzzi A, Chiodini PG, Oppizzi G, Dallabonzana D, Spelt B, Silvestrini F, Rainer E, Horowski R (1983) Medical treatment of pituitary adenomas: effects on hormonal secretion and tumour size. In: Calne DB, Horrowski R, McDonald W, Wuttke W (eds) Lisuride and other dopamine agonists. Raven Press, New York, pp 231–238

62. Lüdecke DK, Saeger W (1980) Characteristics and surgical problems in prolactinomas in comparison with HGH secreting adenomas. Acta Endocrinol [Suppl] 234: 54–55

63. Lüdecke DK, Herrmann HD, Hörmann C, Desaga U, Sager W (1983) Comparison of effects of dopamine agonists and microsurgery in GH and PRL-secreting adenomas. In: Calne DB, Horrowski R, McDonald W, Wuttke W (eds) Lisuride and other dopamine agonists. Raven Press, New York, pp 271–282

64. McGregor AM, Scanlon MF, Hall K, Jall R (1979) Effects of bromocriptine on pituitary tumour size. Br Med J 2: 700–703

65. Melis GB, Gambacciani M, Paoletti AM, Beneventi F, Mais V, Baroldi P, Fioretti P (1987) Dose related prolactin inhibitory effect of the new long-acting dopamine receptor agonist Cabergoline in normal cycling, puerperal and hyperprolactinemic women. J Clin Endocrinol Metab 65: 41–548

66. Molitch ME, Elton RL, Blackwell RE, Caldwell B, Chang RJ, Jaffe R, Joplin G, Robbins RJ, Tyson J, Thorner MO and The Bromocriptine Study Group (1985) Bromocriptine as primary therapy for prolactin-secreting macroadenomas: results of a prospective multicenter study. J Clin Endocrinol Metab 60: 698–705

67. Moriondo P, Travaglini P, Nissim M, Conti A, Faglia G (1985) Bromocriptine treatment of microprolactinomas: Evidence of stable prolactin decrease after drug withdrawal. J Clin Endocrinol Metab 60: 764–772

68. Mosca L, Solcia E, Capella C, Buffa R (1980) Pituitary adenomas: surgical versus past mortem findings today. In: Faglia G, Giovanelli MA, MacLeod RM (eds) Pituitary microadenomas. Academic Press, London, pp 137–149

69. Moses AC, Molitch ME, Sawin CT, Jackson IMD, Biller BJ, Furlanetto R, Reichlin S (1981) Bromocriptine therapy in acromegaly: use on patients resistant to conventional therapy and effect on serum levels of somatomedin-C levels and clinical activity. J Clin Endocrinol Metab 53: 752–756

70. Nortier JWR, Croughs RJM, Thijssen JHH, Schwarz F (1985) Bromocriptine therapy in acromegaly: effects on plasma GH

levels, somatomedin C levels and clinical activity. J Clin Endocrinol 22: 209–217

71. Peillon F, Cesselin F, Bression D, Zygelman N, Brandi AM, Nousbaum A, Mauborgne A (1979) In vitro effect of dopamine and L-dopa on prolactin and growth hormone release from human pituitary adenomas. J Clin Endocrinol Metab 49: 737–744

72. Peillon F, Racadot J, Olivier L, Vila-Porcile E (1980) Microadenoma, structure and function. In: Faglia G, Giovanelli MA, MacLeod RM (eds) Pituitary microadenomas. Academic Press, London, pp 91–105

73. Pieters GFFM, Smals AGH, Kloppenborg PWC (1986) Long-term treatment of acromegaly with the somatostatin analogue SMS 201-995. N Engl J Med 314: 1391

74. Plewe G, Beyer J, Krause U, Neufeld M, del Pozo E (1984) Long-acting and selective suppression of growth hormone secretion by somatostatin analogue SMS 201-995 in acromegaly. Lancet ii: 782–791

75. Plewe G, Schezenmeir J, Nölken G, Kruase U, Beyer J, Kasper H, del Pozo E (1986) Long-term treatment of acromegaly with the somatostatin analogue SMS 201-995 over six months. E Klin Wochenschr 64: 389–395

76. Del Pozo E, Neufeld M, Schlüter K, Tortosa F, Clarenbach P, Bieber E, Wendel L, Nüesch E, Marbach P, Cramer H, Kerp L (1986) Endocrine profile of a long-acting somatostatin analogue SMS 205-995. Study in normal volunteers following subcutaneous administration. Acta Endocrinol 111: 433–439

77. Rasmussen C, Bergh T, Wide L, Brownell J (1987) CV 205-502: A new long-acting drug for inhibition of prolactin hypersecretion. Clin Endocrinol 26: 321–326

78. Reichin S (1983) Somatostatin. N Engl J Med 309: 1495–1501

79. Rodman EF, Molitch ME, Post KD, Biler BJ, Reichlin S (1984) Long-term follow-up of transsphenoidal selective adenomectomy for prolactinomas. JAMA 252: 921–924

80. Rolland R, van der Heijden PFM, Brownell J, van Gennep J (1986) Prolactin lowering effect of CQP 201-403, a new dopamine agonist. In: The 1st International Congress of Neuroendocrinology. Abst. 153. San Francisco, Calif., July, 9–11

81. Saitoh Y, Mori S, Arita N, Hayakawa T, Mogami H, Matsumoto K, Mori H (1986) Cytosuppressive effect of bromocriptine on human prolactinomas: stereological analysis of ultrastructural alterations with special reference to secretory granules. Cancer Res 46: 1507–1512

82. Serri O, Rasio E, Beauregard H, Hardy J, Somma M (1983) Recurrence of hyperprolactinemia after selective transsphenoidal adenomectomy in women with prolactinomas. N Engl J Med 309: 280–283

83. Spada A, Sartorio A, Bassetti M, Pezzo G, Ginnattasio G (1982) In vitro effect of dopamine on growth hormone (GH) release from human GH-secreting pituitary adenomas. J Clin Endocrinol Metab 55: 734–740

84. Thorner MO, Chait A, Aitken M, Benker G, Bloom SR, Mortimer CH, Sanders P, Mason AS, Besser GM (1975) Bromocriptine treatment of acromegaly. Br Med J 1: 299–303

85. Thorner MO, Evans WS, MacLeod RM, Nunley WC, Rogol D, Morris JL, Besser GM (1980) Hyperprolactinemia: Current concepts of management including medical therapy with bromocriptine. In: Goldstein M, Calne D, Lieberman A, Thorner MO (eds) Ergot compounds and brain function: Neuroendocrine and Neuropsychiatric Aspects. Raven Press, New York, pp 165–189

86. Thorner MO, Burdman JA, Calabrese MT, Valdenegro CA, Vance ML, MacLeod RM (1983) Dopamine agonists and prolactinomas: clinical and basic considerations of the mechanism of action. In: Calne DB, Horrowski R, McDonald W, Wuttke W (eds) Lisuride and other dopamine agonists. Raven Press, New York, pp 213–229

87. Tindall GT, McLanahan CS, Christy JH (1978) Transsphenoidal microsurgery for pituitary tumours associated with hyperprolactinemia. J Neurosurg 48: 849–860

88. Tindall GT, Kovacs K, Horvath E, Thorner MD (1982) Human prolactin producing adenomas and bromocriptine: A histological immunocytochemical, ultrastructural and morphometric study. J Clin Endocrinol Metab 55: 1178–1183

89. Tramu G, Beauvillain JC, Mazzuca M, Grenier JL, Fossati P, Christiaens JL (1980) Time dependent evolution of pituitary prolactin adenomas under bromocriptine therapy. In: Derome PJ, Jedynak CP, Peillon F (eds) Pituitary adenomas. Asclepios Publishers, France, p 343

90. Tucker HStG, Lankford HV, Gardner DF, Blachard WG (1980) Persistent defect in regulation of prolactin secretion following successful pituitary tumour removal in women with amenorrhea-galactorrhea syndrome. J Clin Endocrinol Metab 51: 968–977

91. Tucker HSG, Grubb SR, Wigard JP (1982) Galactorrhea-amenorrhea syndrome: follow-up of 45 patients after pituitary tumour removal. Ann Intern Med 94: 302–307

92. Turkalj I, Braun P, Krupp P (1982) Surveillance of bromocriptine in pregnancy. JAMA 247: 1589–1591

93. Wass JAH, Thorner MO, Morris DV, Rees JH, Mason AS, Jones AE, Besser GM (1977) Long-term treatment of acromegaly with bromocriptine. Br Med J 1: 875–878

94. Wass JAH, Moult PJA, Thorner MO, Dacie JE, Charlesworth M, Jones ME, Besser GM (1979) Reduction of pituitary tumour size in patients with prolactinomas and acromegaly treated with bromocriptine with or without radiotherapy. Lancet ii: 66–69

95. Wass JAH, Willians J, Charlesworth M, Kingsley DPE, Halliday M, Doniach I, Rees LH, McDonald WI, Besser GM (1982) Bromocriptine in management of large pituitary tumours. Br Med J 284: 1908–1911

96. Weil C (1986) The safety of bromocriptine in hyperprolactinemic female infertility: a literature review. Curr Med Res Opin 103: 172–195

97. Von Werder K, Eversmann T, Fahlbusch R (1982) Development of hyperprolactinemia in patients with adenomas with or without prior operative treatment. Excerpta Med Int Congr Ser 584: 175–188

98. Von Werder K, Fahlbusch R, Rjosk HK (1983) Macroprolactinomas: Clinical and therapeutical aspects. In: Tolis G, Stefanis C, Mountokalakis T, Labrie F (eds) Prolactin and prolactinomas. Raven Press, New York, pp 415–422

99. Wollesen P, Andersen T, Karle K (1982) Size reduction of extrasellar pituitary tumours during bromocriptine treatment. Ann Intern Med 96: 281–286

100. Winkelman W, Allolio B, Deuss U, Heesen D, Kaulen D, Wileke O (1983) Persistent normoprolactinemia after withdrawal of bromocriptine therapy in patients with prolactinomas. In: 3rd European Workshop on Pituitary Adenomas, Amsterdam, The Netherlands, Abst p 9

Correspondence: Dr. I. Lancranjan, Department of Neuroendocrinology, Clinical Research, Sandoz Ltd., CH-4002 Basel, Switzerland.

Acta Neurochirurgica, Suppl. 47, 86–89 (1990)

Vasopressin and Oxytocin Localization and Putative Functions in the Brain

R. M. Buijs

Netherlands Institute for Brain Research, Amsterdam, The Netherlands

Introduction

In this chapter a description will be given of the widespread occurrence of vasopressin (VP) and oxytocin (OT) neurons in the rodent brain. In addition, an attempt will be made to describe, on the basis of its origin, possible functions of VP and OT in the various parts of the central nervous system (CNS).

The availability of antibodies to VP, OT and neurophysin permitted the localization of these hormones in separate cell bodies in the paraventricular nucleus (PVN) and supraoptic nucleus (SON) and later on in the suprachiasmatic nucleus (SCN) (Vandesande et al. 1974, Swaab et al. 1975). Recently, other sites of VP synthesis were revealed after pretreatment of a rat with cholchicine: in cell bodies of the bed nucleus of the stria terminalis (BST), dorsomedial hypothalamus, medial amygdala and locus coeruleus (Caffé and Van Leeuwen 1983, Van Leeuwen and Caffé 1983).

Improvement of immunocytochemical techniques resulted in the demonstration of extensive VP and OT pathways in the central nervous system. The presence of this central VP innervation suggested that VP is able to reach neuronal structures by other routes than via the general circulation or the cerebrospinal fluid. Indeed, the presence of extensive fiber arborizations and perineuronal structures suggested that these fibers actually terminate in many such areas. Immunoelectronmicroscopy applied to the lateral septum, lateral habenular nucleus, amygdala and nucleus tractus solitarius (NTS) showed that VP fibers indeed terminate synaptically on other neuronal structures (Buijs and Swaab 1979, Voorn and Buijs 1983).

At present, a number of reports have appeared showing the synaptic release of VP or OT in a number of different brain regions (Cooper et al. 1979, Buijs and Van Heerikhuize 1982, Pittman et al. 1984). The conclusion seems justified that VP and OT will be released with a specific stimulus and will act in only those areas where synaptic specializations are present. In addition, the actions of this peptide on neuronal firing in several brain regions, (Zerihun and Harris 1981, Mühlethaler et al. 1982, Joëls and Urban 1982), together with the demonstration of VP and OT binding in the same brain regions (Baskin et al. 1983, Van Leeuwen and Wolters 1983, Biegon et al. 1984), shows that these peptides are able to act as neurotransmitters in these brain regions (Buijs 1983). With a neurotransmitter role of VP or OT basically being established, it is necessary to determine the functional meaning of these transmitters and how and where these functions can be executed. Therefore it became of crucial importance to determine the origin of the VP and OT fibers in the various brain regions. Fig. 1 summarizes the anatomical picture of the organization of the different VP systems that can be found in the rat brain, while Fig. 2 does the same for OT. It will be clear that based on the widespread distribution of VP fibers and the different origin of these fibers, VP is involved in a large number of different brain functions.

For OT the picture seems to be far less complex, most innervations being derived from the PVN, and apart from projections to the circumventricular organs the brain stem and spinal cord are largely reached by OT fibers.

De Vries et al. (1981) detected the presence of a sexually dimorphic innervation of VP fibers in the lateral septum. From postnatal day 12 onwards, a male rat had a much denser VP innervation in the lateral septum than a female rat. Moreover, the presence of gonadal hormones, either testosterone or oestradiol, appeared to be of crucial importance for the mainte-

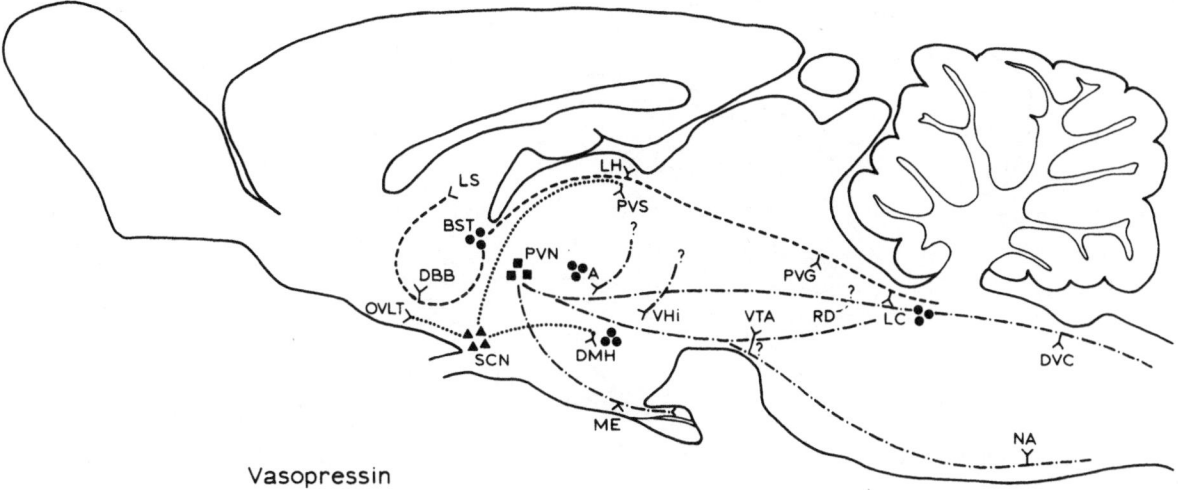

Vasopressin

Fig. 1. Sagittal scheme of the rat brain representing the major vasopressinergic pathways with their most likely origin. The pathways of the paraventricular nucleus are indicated by (— · — · —). Pathways of the suprachiasmatic nucleus are indicated by (· · · · · ·). The vasopressin cell groups found after cholchicine treatment are indicated by black dots, while the pathways of the bed nucleus of the stria terminalis are indicated by (— — — — —). *A* amygdala; *DBB* diagonal band of Broca; *DHM* dorsomedial nucleus of the hypothalamus; *DVC* dorsal vagal complex; *LC* locus coeruleus; *LH* lateral habenula; *LS* lateral septum; *ME* median eminence; *NA* nucleus ambiguus; *OVLT* organum vasculosum of the laminae terminalis; *PVG* paraventricular grey; *PVS* periventricular nucleus; *RD* dorsal raphe nucleus; *SCN* substantia nigra; *VHi* ventral hippocampus; *VTA* ventral tegmental area

nance of the VP fiber innervation of, *e.g.*, the lateral septum (De Vries *et al.* 1985). Gonadectomy of adult male or female rats resulted in a gradual decrease in VP fiber density over a period of 15 weeks to a point where hardly any fibers were found (De Vries *et al.* 1984). These results indicate that apart from the possibility of dividing the extrahypothalamic pathways on the basis of their origin, it is also possible to divide them on the basis of their sensitivity to gonadal hormones.

At present, the morphology and origin of the various VP systems is well established and it may be possible to make a few assumptions on the putative functions of VP in the brain.

Each of the sites where VP is produced will be influenced by different stimuli, a spatially controlled release of VP. Therefore the approach of the present chapter is to take into account the source of VP fibers when it comes to distinguishing between the different VP systems of the brain.

Paraventricular Nucleus

The PVN supplies the largest contingent of VP and OT fibers, the latter predominating, in the caudal brain, viz., brain stem and spinal cord (Buijs 1978, Swanson and McKellar 1979). The OT projection to the A1 and A2 areas and to the intermediolateral column in the

spinal cord have been the most eye-catching of the PVN OT projections, suggesting that these peptides from the PVN are involved in the control of autonomous functions. It has been shown that especially A1 (noradrenergic) neurons (Sawchenko and Swanson 1981, Sladek and Zimmerman 1982) densely innervate the magnocellular VP and OT containing neurons in the SON and PVN. Day and Renaud (1984) have shown by a combination of lesioning with 6-hydroxy-dopamine and stimulation of the A1 area that this noradrenergic innervation provides a stimulatory input for the VP containing neurons of the SON and PVN, thus promoting the neurohypophyseal release of VP into the bloodstream. The innervation of A1 and A2 neurons by OT containing fibers suggests that OT as a neurotransmitter can be involved in, for example, the control of the release of VP from the HNS into the general circulation. Infusions of small amounts of OT in this area indeed induce elevated levels of vasopressin in the general circulation (Hermes, unpublished observation).

Suprachiasmatic Nucleus

The SCN, which is the major endogenous pacemaker, functions as the major clock in synchronizing the circadian rhythms (Stephan and Zucker 1972, Moore 1979). It seems only logical to assume that VP from the SCN is involved, one way or another, in this syn-

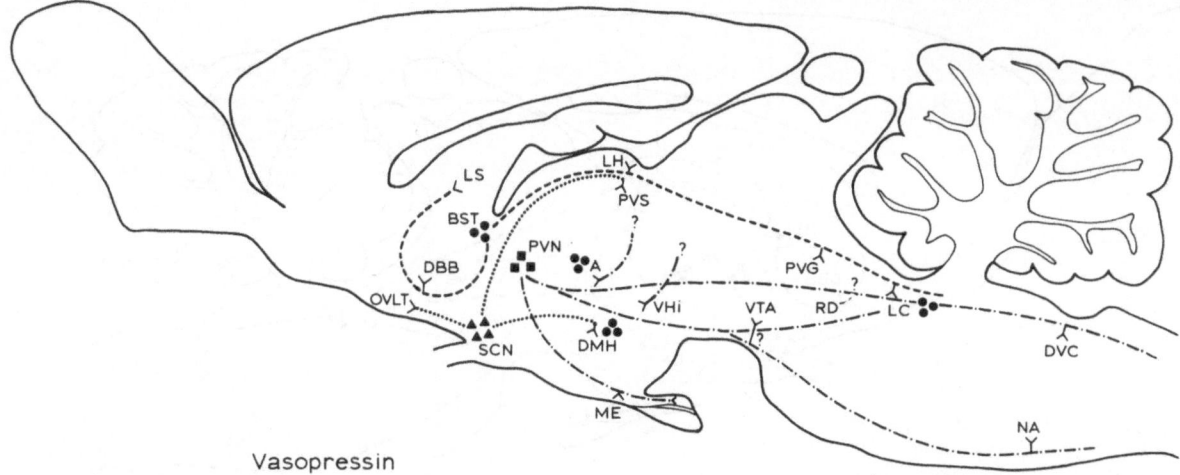

Fig. 2. Sagittal scheme of the rat brain representing the major oxytocinergic pathways originating largely from the paraventricular nucleus. For abbreviations used, see Fig. 1

chronizing function. The fact that a VP-deficient Brattleboro rat, which lacks VP also in the SCN, seems to have a normal sleep-wake rhythm (Peterson *et al.* 1980), does not necessarily lead to the conclusion that VP is of no importance to the circadian synchronization of certain behaviors. These mutant rats are likely to have compensatory systems that, at least partially, take over the function of VP.

In the CSF, VP has a pronounced day-night rhythm that seems to be dependent on the activity of the SCN (Reppert *et al.* 1982, Schwartz *et al.* 1983). Södersten *et al.* (1983, 1985) have shown that VP injected into the CSF during low VP night levels inhibits lordosis behavior in ovariectomized female rats, while an antagonist injected during the day period enhance lordosis behavior. Their observations led them to the conclusion that VP inhibits lordosis behavior, and to the suggestion that the SCN is involved in the regulation of this behavior.

Bed Nucleus for the Stria Terminalis and Medial Amygdala

The reason for discussing these nuclei together is that they are both influenced by gonadal steroids and probably both project to the same areas (Holstege *et al.* 1985). The sexually dimorphic innervation, and the fact that replacement of oestrogen and testosterone restores the original VP innervation (in gonadectomized females and males respectively; De Vries *et al.* 1981, 1984) suggests that VP in these limbic brain regions is involved in the regulation of sex-linked functions. This does not imply, however, that VP is involved in the regulation

of reproductive behavior, for a number of other behaviors such as feeding, drinking, temperature regulation and aggressive behavior is sexually dimorphic as well (Goy and McEwen 1980, Beatty 1984). Thus, the sexually dimorphic VP innervation may be involved in non-reproductive behavior-linked sexually dimorphic functions.

References

Baskin DG, Petracca FM, Dorsa DM (1983) Autoradiographic localization of specific binding sites for [3H]-arginine 8-vasopressin in the septum of the rat brain with tritium-sensitive film. Eur J Pharmacol 90: 155–157

Beatty W (1984) Hormonal organization of sex differences in play fighting and spatial behavior. Progr Brain Res 61: 386–397

Biegon A, Terlou M, Voorhuis ThD, De Kloet ER (1984) Arginine-vasopressin binding sites in the rat brain: A quantitative autoradiographic study. Neurosci Lett 44: 229–234

Buijs RM (1978) Intra- and extrahypothalamic vasopressin and oxytocin pathways in the rat. Pathways to the limbic system, medulla oblongata and spinal cord. Cell Tiss Res 192: 423–435

— (1983) Vasopressin and oxytocin—their role in neurotransmission. Pharmacol Ther 22: 127–141

— Swaab DF (1979) Immunoelectronmicroscopical demonstration of vasopressin and oxytocin in the limbic system of the rat. Cell Tiss Res 204: 355–365

— Van Heerikhuize JJ (1982) Vasopressin and oxytocin release in the brain: A synaptic event. Brain Res 252: 71–76

Burbach JPH, Wang X-C, Ten Haaf JA, De Wied D (1984) Substances resembling C-terminal vasopressin fragments are present in the brain but not in the pituitary. Brain Res 306: 384–387

Caffé AR, VAn Leeuwen FW (1983) Vasopressin-immunoreactive cells in the dorsomedial hypothalamic region, medial amygdaloid nucleus and locus coeruleus of the rat. Cell Tiss Res 233: 23–33

Cooper KE, Kasting NW, Lederis K, Veale WL (1979) Evidence supporting a role for endogenous vasopressin in natural suppression of fever in the sleep. J Physiol (Lond) 295: 33–45

Day TA, Renaud LP (1984) Electrophysiological evidence that no-radrenergic afferents selectively facilitate the activity of supraoptic vasopressin neurons. Brain Res 303: 233–240

De Vries GJ, Buijs RM, Swaab DF (1981) Ontogeny of the vaso-pressinergic neurons of the suprachiasmatic nucleus and their extrahypothalamic projections in the rat brain—presence of a sex difference in the lateral septum. Brain Res 218: 67–78

— — Sluiter AA (1984) Gonadal hormone actions on the mor-phology of the vasopressinergic innervation of the adult rat brain. Brain Res 298: 141–145

— — Van Leeuwen FW, Caffé AR, Swaab DF (1985) The vaso-pressinergic innervation of the brain in normal and castrated rats. J Comp Neurol 233: 236–254

Goy RW, McEwen BS (1980) Sexual differentiation of the brain. MIT Press, Boston

Hermes MLHJ, Unpublished observation

Holstege G, Meiners L, Tan K (1985) Projections of the bed nucleus of the stria terminalis to the mesencephalon, pons, and medulla oblongata in the cat. Exp Brain Res 58: 379–391

Joëls M, Urban JJA (1982) The effect of microiontophoretically applied vasopressin and oxytocin on single neurones in the sep-tum. Neurosci Lett 33: 79–84

Moore RY (1979) The anatomy of central neural mechanisms reg-ulating endocrine rhythms. In: Krieger DT (ed) Endocrine rhythms. Raven Press, New York, pp 63–78

Mühlethaler M, Dreifuss JJ, Gahwiler BH (1982) Vasopressin causes excitation of hippocampal neurons. Nature 296: 749–751

Peterson GM, Watkins WB, Moore RY (1980) The suprachiasmatic hypothalamic nuclei of the rat. VI. Vasopressin neurons and circadian rhythmicity. Behav Neur Biol 29: 236–245

Pittman QJ, Riphagen CL, Lederis K (1984) Release of immunoas-sayable neurohypophyseal peptides from rat spinal cord, in vivo. Brain Res 300: 321–326

Reppert SM, Coleman RJ, Heath HW, Keutmann HT (1982) Cir-cadian properties of vasopressin and melatonin rhythms in cat cerebrospinal fluid. Am J Physiol 243: 489–498

Sawchenko PE, Swanson LW (1981) Central noradrenergic pathways for the integration of hypothalamic neuroendocrine and auto-nomic responses. Science 214: 685–687

Schwartz WJ, Coleman RJ, Reppert SM (1983) A daily vasopressin rhythm in rat cerebrospinal fluid. Brain Res 263: 105–112

Sladek JR Jr, Zimmerman EA (1982) Simultaneous monoamine histofluorescence and neuropeptide immunocytochemistry VI. Catecholamine innervation of vasopressin and oxytocin neurons in the rhesus monkey hypothalamus. Brain Res Bull 9: 431–440

Södersten P, Henning M, Melin P, Lundin S (1983) Vasopressin alters female sexual behaviour by acting on the brain indepen-dently of alterations in blood pressure. Nature 301: 608–610

— De Vries GJ, Buijs RM, Melin P (1985) A daily rhythm in be-havioral vasopressin sensitivity and brain vasopressin concen-trations. Neurosci Lett 58: 37–41

Stephan FK, Zucker I (1972) Circadian rhythms in drinking behavior and locomotor activity of rats are eliminated by hypothalamic lesions. Proc Natl Acad Sci USA 69: 1583–1587

Swaab DF, Pool CW, Nijveldt F (1975) Immunofluorescence of vasopressin and oxytocin in the rat hypothalamo-neurohypo-physeal system. J Neural Transm 36: 195–215

Swanson LW, McKellar S (1979) The distribution of oxytocin- and neurophysin-stained fibers in the spinal cord of the rat and mon-key. J Comp Neurol 188: 87–106

Vandesande F, De Mey J, Dierickx K (1974) Identification of neu-rophysin producing cells. I. The origin of the neurophysin-like substance-containing nerve fibres of the external region of the median eminence of the rat. Cell Tiss Res 151: 187–200

Van Leeuwen FW, Caffé AR (1983) Immunoreactive vasopressin cell bodies in the rat bed nucleus of the stria terminalis. Cell Tiss Res 228: 525–534

— Wolters P (1983) Light microscopic autoradiographic localiza-tion of [3H]arginine vasopressin binding sites in the rat brain and kidney. Neurosci Lett 41: 61–66

Voorn P, Buijs RM (1983) An immuno-electronmicroscopical study comparing vasopressin, oxytocin, substance P and enkephalin containing nerve terminals in the nucleus of the solitary tract of the rat. Brain Res 270: 169–173

Zerihun L, Harris M (1981) Electrophysiological identification of neurones of paraventricular nucleus sending axons to both the neurohypophysis and the medulla in the rat. Neurosci Lett 23: 157–160

Correspondence: R. M. Buijs, M.D., Netherlands Institute for Brain Research, Meibergdreef 33, 1105 AZ Amsterdam ZO, The Netherlands.

Acta Neurochirurgica, Suppl. 47, 90–94 (1990)

Central Nervous System Control of Fluid Balance: Physiology and Pathology

S. Lightman

Charing Cross and Westminster Medical School, Westminster Hospital, Neuroendocrinology Unit, London, U.K.

Introduction

The water content of the body is divided between extracellular and intracellular compartments. The 33% of body water in the extracellular fluid is very finely regulated and even following major physiological changes such as exercise, drinking or eating, plasma osmolality remains within a narrow range of about 286–294 mmol/kg. The pre-eminent mechanism controlling body water content is the hormone arginine vasopressin (AVP) which is synthesized in the supraoptic and paraventricular nuclei of the hypothalamus, and acts on the kidney to reduce free water clearance.

Vasopressin secretion can be activated by osmotic and cardiovascular stimuli. The osmoreceptors are predominantly located in the hypothalamus itself, probably in the anteroventral part of the third ventricle whence they are thought to project directly to the vasopressin containing magnocellular cells of the supraoptic and paraventricular nuclei (Lightman and Everitt 1986). Afferent fibres from the cardiovascular receptors in the thorax travel in the IX and X cranial

nerves to synapse in the nucleus of the tractus solitarius (NTS) in the dorsomedial medulla. From the nucleus of the tractus solitarius two noradrenergic pathways project to the hypothalamus (Sawchenko and Swanson 1982). A major nonadrenergic pathway connects the NTS to the A1 medullary noradrenergic cell group which then projects directly to the hypothalamus; whilst a second noradrenergic pathway carries fibres from the A2 cell group within the NTS itself, the A6 (locus coeruleus) in the pons and some dorsal projections from A1. This pathway innervates most of the forebrain and also contributes towards hypothalamic catecholamine content (Fig. 1). It is generally agreed that the A1 projections run primarily within the ventral noradrenergic bundle of fibres, and the A2, A6 and some A1 projections in the dorsal noradrenergic bundle. In addition to the neuroanatomical description of these brainstem noradrenergic pathways there is also much information on the effect of noradrenaline on the firing of hypothalamic vasopressin cells, and although the situation is not completely clear it seems noradrenaline does indeed facilitate the firing of vasopressin cells (Day et al. 1984).

We have been particularly interested in the mechanisms controlling vasopressin secretion in man and the effect of pathological lesions on its control. We have studied three major mechanisms which modulate the release of vasopressin—osmotic stimuli, cardiovascular stimuli and oropharyngeal stimuli.

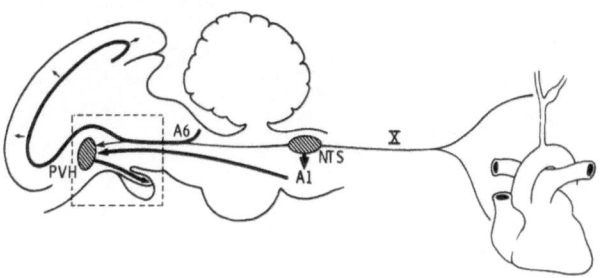

Fig. 1. Schematic representation of the catecholamine projections from the brainstem to the paraventricular nucleus of the hypothalamus (*PVH*) and of neurosecretory fibres from the PVH to the pituitary neural lobe. The dorsal noradrenergic bundle runs between A6 (the locus coeruleus) and the hypothalamus, the ventral noradrenergic bundle between A1 and the hypothalamus

Osmotic Stimuli

In normal subjects there is a close relationship between plasma omolality and plasma AVP concentrations (Lightman and Everitt 1986), as plasma osmolality increases so does the plasma concentration of vasopres-

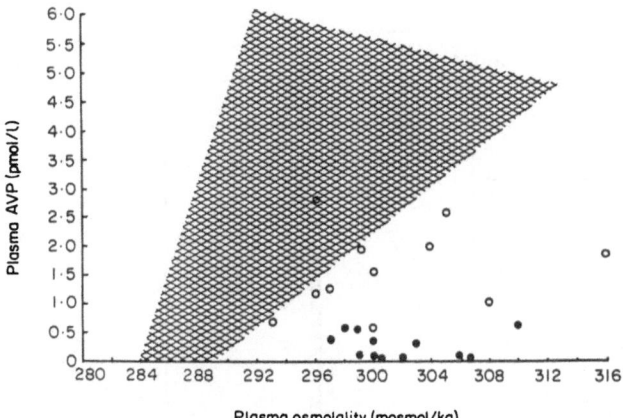

Fig. 2. Response of plasma vasopressin (AVP) during infusion of 0.85 M saline in two subjects with selective osmoreceptor dysfunction. Normal results should fall within the cross-hatched area

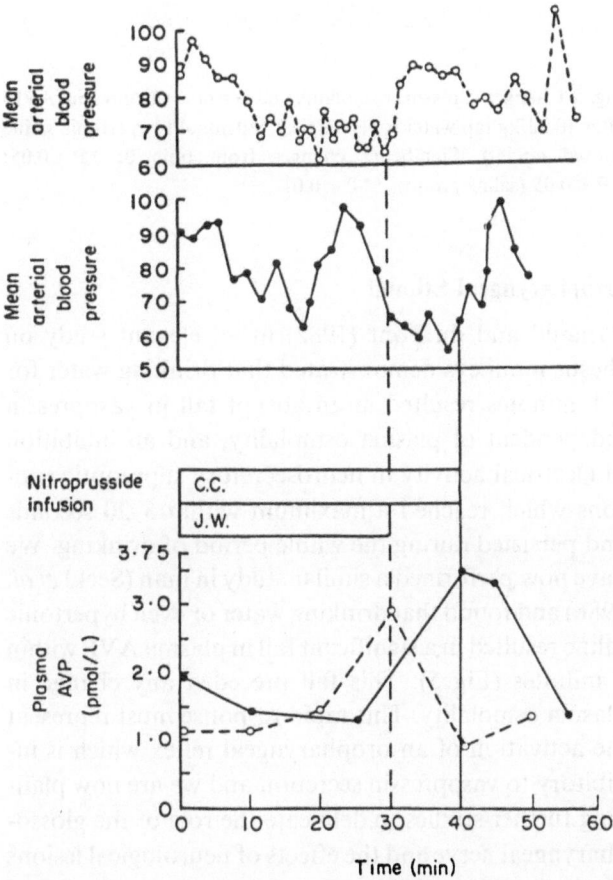

Fig. 3. Plasma AVP and mean arterial blood pressure during nitroprusside infusion in 2 subjects with selective osmoreceptor dysfunction

sin. The site of the osmoreceptors which respond to these changes in osmolality has been a subject of some dispute. It is, however, generally agreed that the osmosensitive areas of hypothalamus reside in the anteroventral region of the third ventricle in the rat, an

area which includes the organum vasculosum of the lamina terminalis found in larger animals. Lesions of this area in the rat (Sladek and Johnson 1983), or dog (Thrasher *et al.* 1982) result in a loss of vasopressin response to osmotic stimuli. Interestingly, they maintain their vasopressin response to cardiovascular stimuli so that the brainstem pathway and final common hypothalamo-neurohypophyseal pathways must remain intact.

Similar abnormalities can occur in man and are usually associated with an abnormality of thirst and—not surprisingly—hypernatraemia. We have recently described two such patients (Dunger *et al.* 1985) and their lack of response to hypertonic saline can be seen clearly in Fig. 2. Like dogs with lesions of the organum vasculosum of the lamina terminalis, these subjects still maintain their response to a cardiovascular stimulus (Fig. 3) which in our study was elicited by the use of controlled hypotension with intravenous sodium nitroprusside.

Cardiovascular Stimuli

A demonstration of the vasopressin response to a cardiovascular stimulus has already been demonstrated in Fig. 3. We have studied the brainstem pathways which mediate these responses and have found that in the rat, noradrenergic pathways which run in the dorsal noradrenergic bundle (Fig. 1) play a major role in the cardiovascular control of vasopressin secretion (Lightman *et al.* 1984). This effect is specific for the cardiovascular control of vasopressin since lesions of the dorsal noradrenergic bundle do not effect the vasopressin response to an osmotic stimulus.

Although there is no clear model of discrete bilateral lesions of the noradrenergic pathways in the human brainstem, there is one condition associated with a loss of brainstem amines. In the Shy-Drager syndrome not only is there histological evidence for neuronal loss in the locus coeruleus (Oppenheimer 1983), there is also neurochemical evidence for a loss of catecholamines both in the brainstem and hypothalamus (Spokes *et al.* 1979). It is clearly of interest to know whether this loss of brainstem and hypothalamic catecholamines is associated with any alteration in the cardiovascular or osmotic control of vasopressin secretion.

We have now been able to compare the vasopressin response to the osmotic stimulus of intravenous hypertonic saline and the cardiovascular stimulus of upright tilt in three groups of subjects: Normal controls, patients with the Shy-Drager syndrome and patients

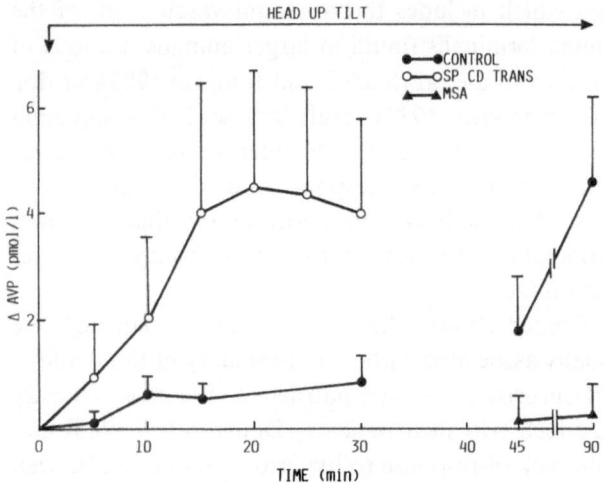

Fig. 4. Changes in plasma vasopressin (AVP) during head-up tilt in controls, Shy-Dragers (MSA) and Tetraplegics (SPCD Trans)

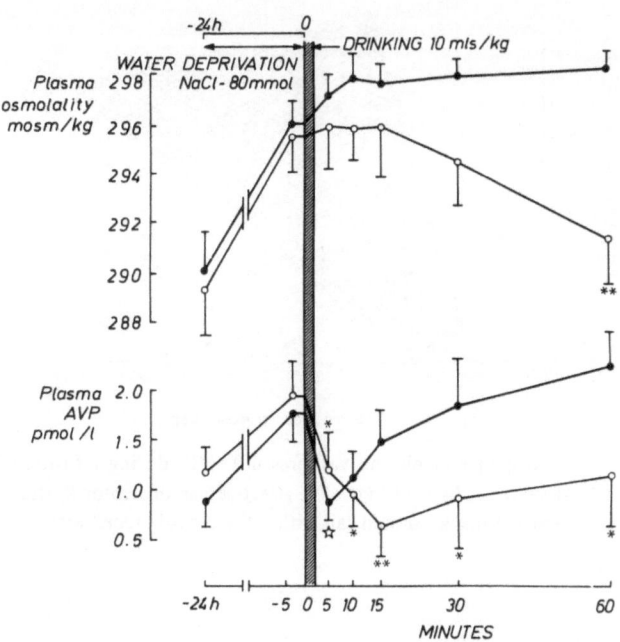

Fig. 5. Changes in plasma osmolality and arginine vasopressin (AVP) after 10 ml/kg tap water (open circles) 180 mmol/l hypertonic saline (closed circles). Significant changes from time 0: *P < 0.05; ☆P < 0.05 (saline group); **P < 0.01

with mid-cervical cord lesions (these act as controls for the autonomic failure and postural hypotension found in the patients with the Shy-Drager syndrome) (Puritz et al. 1983, Poole et al. 1987).

The vasopressin response to hypertonic saline was normal in all three groups of subjects. We can therefore deduce that the vasopressin response to an osmotic stimulus is intact and not inhibited by the loss of brain-stem aminergic pathways in the Shy-Drager syndrome nor are they effected by cervical cord transection.

In contrast to the normal response to hypertonic saline there is a marked difference in the responses of the three groups to the cardiovascular stimuli of head-up tilt (Fig. 4). Normal subjects show a significant rise in plasma vasopressin following tilt. The additional stimulus of postural hypotension in the subjects with mid-cervical cord transection was undoubtedly the cause of the greater vasopressin response to tilt in these subjects. In spite of a very similar hypotensive response to tilt however, the vasopressin response in the group with Shy-Drager syndrome was only 10% of that found in their control group. It is clear there is a marked similarity between the neuroendocrine abnormalities manifest in subjects with Shy-Drager syndrome and in our rats with lesions of the dorsal noradrenergic bundle. Both maintain a normal vasopressin response to os-motic stimuli but have a severely blunted response to a cardiovascular stimulus. This suggests that in man, like the rat, ascending catecholamine pathways are im-portant in mediating the vasopressin response to car-diovascular stimuli.

Oropharyngeal Stimuli

Arnauld and du Pont (1982) in an elegant study on rhesus monkeys demonstrated that drinking water for 5–8 minutes resulted in an abrupt fall in vasopressin independent of plasma osmolality, and an inhibition of electrical activity in neurosecretory supraoptic neu-rons which reached a maximum within 5–20 seconds and persisted during the whole period of drinking. We have now performed a similar study in man (Seckl et al. 1986) and found that drinking water or even hypertonic saline resulted in a significant fall in plasma AVP within 5 minutes (Fig. 5). This fall preceded any change in plasma osmolality. This rapid response must represent the activation of an oropharyngeal reflex which is in-hibitory to vasopressin secretion and we are now plan-ning further studies to delineate the role of the glosso-pharyngeal nerve and the effects of neurological lesions on the integrity of this reflex.

Post-Operative Diabetes Insipidus

Diabetes insipidus may occur following surgery to the pituitary or hypothalamus and exhibits one of three patterns (Randall et al. 1960, Verbalis et al. 1985). These are illustrated in Fig. 6:

1. Transient DI: The polyuria is of abrupt onset usually within the first 24 hours of surgery and resolves

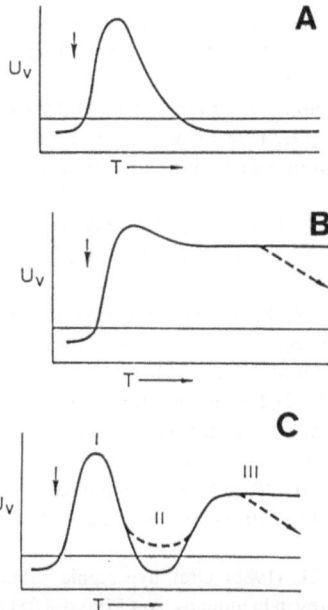

Fig. 6. Three patterns of Diabetes insipidus following hypothalamic or pituitary surgery. U_v Urinary volume, T time. Details in text

over several days. This form is particularly associated with intrasellar surgery.

2. Prolonged or permanent DI: Again the onset is early and rapid but the disorder persists for weeks or may be permanent. This pattern is associated with more proximal damage to the neurohypophyseal stalk or the hypothalamus.

3. The triple response: following a similar early onset of diabetes insipidus there is a transient remission from polyuria (interphase) lasting a few days until prolonged DI occurs (Ikkos *et al.* 1955, Verbalis *et al.* 1985). The resolution phase may be only partial or may even involve a frank syndrome of inappropriate antidiuresis (Lipsett *et al.* 1956) and is thought to be due

to non-specific release of arginine vasopressin from damaged neurohypophyseal tissue degenerating *in situ*.

Although these phases of post-operative DI would be expected to correlate with inadequate plasma concentrations of AVP, there has been little experimental evidence for this. We have now investigated the plasma concentration of AVP, its carrier protein neurophysin 1 and oxytocin in children undergoing pituitary and suprasellar surgery (Seckl *et al.* 1987). DI occurred in approximately 80% of our patients with an abrupt onset within 24 hours of surgery. At this time, however, plasma AVP concentrations were not depressed but were in fact relatively high, at levels usually associated with maximum antidiuresis (Fig. 7). These high levels of AVP and their subsequent fall over the next 24 hours were confirmed by parallel changes in neurophysin 1. Sine HPLC analysis revealed that vasopressin immunoreactivity to be predominantly genuine vasopressin we must conclude that for some reason the kidneys are unresponsive to vasopressin. This is unlikely to be due to non-specific effects, such as changes in sympathetic tone following surgery, since cranial DI does not commonly occur after other neurosurgical procedures. I should like to speculate that the hypothalamic damage may result in the release of other biologically active agents, perhaps related to the hypothalamic natriuretic compound of de Wardener (Clarkson *et al.* 1974) which themselves have a diuretic or antivasopressinergic action.

Conclusion

We are at an early stage in elucidating pathways which mediate the control of vasopressin secretion. We have clear evidence for three major afferent pathways from osmoreceptors, cardiovascular receptors and orophar-

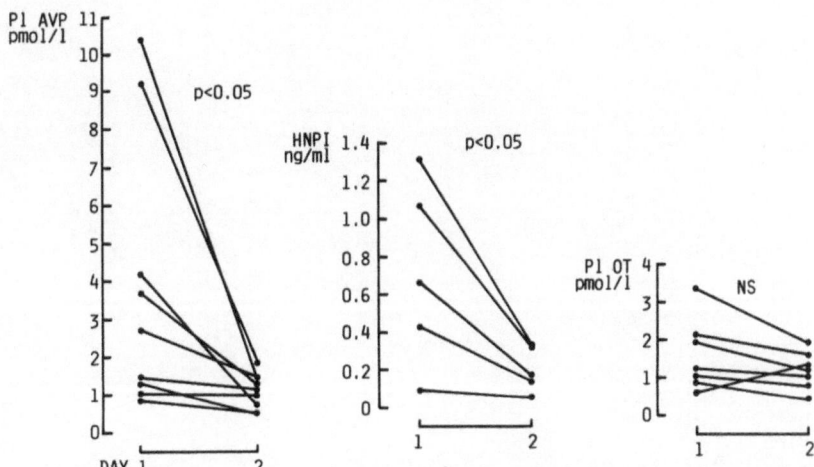

Fig. 7. Changes in plasma vasopressin (*Pl AVP*), human neurophysin I (*HNPI*) and plasma oxytocin (*Pl OT*) during the first 2 days following the onset of postoperative diabetes insipidus in children

yngeal receptors and have demonstrated the results of disruption of some of these pathways in specific neurological or neurosurgical conditions. Post-operative diabetes insipidus is still poorly understood, but we are now in a position to study the aetiology of this condition in more detail to realise a more rational attitude to its management.

References

Arnauld E, du Pont J (1982) Vasopressin release and firing of supraoptic neurosecretory neurones during drinking in the dehydrated monkey. Pflügers Arch 394: 195–201

Clarkson EM, Koutsaimanis KG, Davidman M, Dubois M, Penn WP, de Wardener HE (1974) The effect of brain extracts on urinary sodium excretion of the rat and the intracellular sodium concentration of renal tubule fragments. Clin Sci Mol Med 47: 201–213

Day TA, Ferguson AV, Renaud LP (1984) Facilitatory influence of noradrenergic afferents on the excitability of rat paraventricular nucleus neurosecretory cells. J Physiol 355: 237–249

Dunger DB, Lightman SL, Williams M, Preece MA, Grant DB (1985) Lack of thirst, osmoreceptor dysfunction, early puberty and abnormally aggressive behaviour in two boys. Clin Endocrinol 22: 469–478

Ikkos D, Luft R, Olivecrona H (1955) Hypophysectomy in man: effect on water excretion during the first two postoperative months. J Clin Endocrinol Metab 15: 553–567

Lightman SL, Everitt BJ, Todd K (1984) Ascending noradrenergic projections from the brainstem; evidence for a major role in the regulation of blood pressure and vasopressin secretion. Exp Brain Res 55: 145–151

— — (1986) Water excretion. In: Lightman SL, Everitt BJ (eds) Neuroendocrinology. Blackwell Scientific Publications, Oxford, pp 197–206

Lipsett MB, Maclean JP, West CD, Li MC, Pearson OH (1956) An analysis of the polyuria induced by hypophysectomy in man. J Clin Endocrinol Metab 16: 183–195

Oppenheimer D (1983) Neuropathology of progressive autonomic failure. In: Bannister R (ed) Autonomic failure. Oxford Medical Publications, Oxford, pp 267–283

Poole CJM, Williams TDM, Lightman SL, Frankel HL (1987) Neuroendocrine control of vasopressin secretion and its effect on blood pressure in subjects with spinal cord transection. Brain 110: 727–735

Puritz R, Lightman SL, Wilcox CS, Forsling M, Bannister R (1983) Blood pressure and vasopressin in progressive autonomic failure. Brain 106: 503–511

Randall RV, Clark EC, Dodge HW, Love JG (1960) Polyuria after operation for tumors in the region of the hypophysis and hypothalamus. J Clin Endocrinol Metab 20: 1614–1621

Sawchenko PE, Swanson LW (1982) The organisation of noradrenergic pathways from the brainstem to the paraventricular and supraoptic nuclei in the rat. Brain Res Rev 4: 275–325

Seckl JR, Dunger DB, Lightman SL (1987) Neurohypophyseal peptide function during early postoperative diabetes insipidus. Brain 110: 737–746

— Williams TDM, Lightman SL (1986) Oral hypertonic saline causes transient fall of vasopressin in humans. Am J Physiol 2511 R214–217

Sladek CD, Johnson AK (1983) Effects of anteroventral third ventricle lesions on vasopressin release by organ-cultured hypothalamo-neurohypophyseal explants. Neuroendocrinology 37: 78–84

Spokes EG, Bannister R, Oppenheimer DR (1979) Multiple system atrophy with autonomic failure: Clinical, histological and neurochemical observations on four cases. J Neurol Sci 43: 59–82

Thrasher TN, Keil LC, Ramsay DJ (1982) Lesions of the organum vasculosum of the lamina terminalis (OVLT) attenuate osmotically induced drinking and vasopressin secretion in the dog. Endocrinology 100: 1837–1839

Verbalis JG, Robinson AG, Moses AM (1985) Postoperative and post-traumatic diabetes insipidus. In: Czernichow P, Robinson AG (eds) Diabetes insipidus in man. Frontiers of hormone research, Vol 13. Basel, Karger, pp 242–265

Correspondence: Stafford Lightman, MA, PhD, FRCP, Professor of Clinical Neuroendocrinology, Charing Cross and Westminster Medical School, Westminster Hospital, London, SW1P 2AP, U.K.

Acta Neurochirurgica, Suppl. 47, 95–101 (1990)

Fluid Balance Disturbances in Neurosurgical Patients: Physiological Basis and Definitions

V. Walker

Chemical Pathology and Human Metabolism, Southampton University Medical School, Southampton, England

Normally, neural, hormonal, haemodynamic, and renal mechanisms all function in a highly integrated way to preserve sodium and water homeostasis. There are two major objectives. The first is to keep the *concentration* of sodium in the extracellular fluid (ECF) within very tight limits. Together with its associated anions, sodium constitutes more than ninety per cent of the total solute in the ECF, and it controls the distribution of water between the cells and the extracellular space. Large deviations in ECF sodium concentration from normal cause the cells to shrink or swell. This can have disastrous effects, particularly on brain function. The body protects the sodium concentration constantly by finely adjusting the water content of the ECF. This is achieved through the secretion and action of antidiuretic hormone (ADH) to control water loss from the kidney. The thirst mechanism helps by controlling the fluid intake to some extent. The second objective is to keep the total sodium *content* of the ECF within narrow confines, and thereby to maintain a normal ECF volume. Given that sodium is the major cation of the ECF and that the body adjusts the water around it to maintain a normal sodium concentration, then the total number of sodium ions in the ECF will determine the ECF volume. Large deviations in the ECF sodium content from normal cause fluctuations in the circulating blood volume. Both volume contraction and expansion can have disastrous consequences on brain function, particularly in the presence of brain damage. Normally ECF sodium is regulated by many closely coordinated mechanisms which adjust the amount of sodium lost through the kidneys (Schrier and Anderson 1980). Minor alterations in renal tubular reabsorption can have a profound effect on sodium balance, since the filtered

load of sodium is enormous in relation to the amount excreted (Leader, Lancet 1984).

Fluid balance disturbances are common among neurological patients and are frequently multifactorial in origin. Fluid adjustments occur as part of the normal metabolic response to trauma, haemorrhage, or surgery. Superimposed on these changes may be a neuroendocrine disturbance due to structural brain damage. Therapeutic manœuvres such as the infusion of mannitol, hypotonic or hypertonic solutions may add to the problems. It is essential to consider all aspects of sodium and water homeostasis in order to identify individual factors contributing to a disturbance. Then appropriate therapy can be given. Hyponatraemia in neurosurgical patients has often been attributed to a cerebral salt wasting state, or to inappropriate secretion of antidiuretic hormone (ADH). Whilst these disorders undoubtedly occur (Cort and Yale 1954, Lester and Nelson 1981), recent reports have questioned the accuracy of these diagnoses in some patients (Nelson *et al.* 1981, Bouzarth and Shenkin 1982). Hyponatraemia is probably often iatrogenic.

Factors in the Regulation of Sodium and Water Balance. Arginine Vasopressin (AVP)

AVP is the major mammalian antidiuretic hormone. Its main function is to regulate water balance by promoting reabsorption of water from the renal collecting ducts as they pass through the hyperosmolar gradient in the medulla of the kidney. At least four physiological stimuli release AVP into the blood stream by independent mechanisms in man (Zerbe *et al.* 1981): osmotic, haemodynamic (hypotension and/or hypovolaemia), nausea and emesis, and hypoglycaemia. Of these, osmotic stimulation is normally the most important for fine control of secretion and maintenance of water balance.

Osmoregulation: Changes in plasma osmotic pressure are "sensed" by osmosensitive cells in the brain which relay to AVP-producing cells in the supraoptic nucleus (SON) and paraventricular nucleus (PVN) of the hypothalamus. These osmoreceptors probably lie outside the blood brain barrier, perhaps in the region anterior and ventral to the third ventricle (Lightman and Everitt 1986). An increase in the ECF concentration of sodium ions, which are effectively excluded by the cell membranes, causes dehydration of the osmoreceptor cells. This stimulus is transmitted to SON and PVN, and triggers the release of AVP from the pituitary. Other solutes which are impermeable to cell membranes, for example mannitol, also stimulate the osmoreceptors. Solutes such as urea and glucose, which readily enter cells and do not therefore produce an osmotic gradient, have little or no effect. Osmoregulation is normally extraordinarily sensitive. For every unit rise in plasma osmolality, plasma AVP rose on average by 0.63 pmol/l, and a change in plasma osmolality of only one per cent altered AVP by 1.8 pmol/l (Zerbe *et al.* 1981). A threshold of the osmostat can be defined, generally at a plasma osmolality of around 280–284 mosmol/kg water, at which plasma AVP starts to increase. Below this, secretion is suppressed. Maximum urinary concentration is achieved at a plasma osmolality of around 295 mosml/kg water. The theshold for thirst sensation is around 299 mosmol/kg water but varies widely (Reviewed by Zerbe *et al.* 1981).

Cardiovascular regulation is mediated by high pressure receptors (baroreceptors) in the carotid body and aortic arch, which respond to arterial blood pressure, and by low pressure receptors in the atria of the heart, which respond to changes in blood volume. The afferents travel via the IX and X cranial nerves (Lightman and Everitt 1986). Relatively large stimuli are required in healthy adults to increase plasma AVP: a fall in mean arterial pressure exceeding 5 per cent, or of blood volume of 10–20 per cent (Edwards 1979). Both high and low pressure receptors are stimulated if hypotension accompanies vasodilatation, as in syncope. Once the stimulatory threshold is reached, there is a geometric rise in plasma AVP (Schrier 1980) to concentrations much higher than those achieved with maximal osmotic stimulation, and this over-rides the normal suppression in response to hypotonicity. Under con-

ditions of hypotension or hypovolaemia, the threshold "set" of the osmoreceptor is lowered.

Nausea and emesis lead to large increases in plasma AVP by an unknown mechanism. *Insulin-induced hypoglycaemia* leads to a 2–4-fold increase in man, but the physiological significance of this is uncertain. The widely held belief that pain and other noxious stimuli stimulate AVP release has been questioned (Zerbe *et al.* 1981). Other non-physiological stimuli may be important in neurosurgical patients, for example a sudden increase in intracranial pressure (Rap and Chwalbinska-Moneta 1978). Of anaesthetic agents in common use, halothane and chloralose produced modest increases, and trilene more marked increases, in rats. Phenobarbitone and althesin (alphaxolone/alphadolone) were without effect. Morphine in high doses was stimulatory (Aziz and Forsling 1979). Positive pressure ventilation increased release (Haas and Glick 1978). Among drugs, atropine and chlorpromazine inhibit AVP release. Chlorpropamide and carbamazepine potentiate, and lithium and demethylchlortetracycline inhibit its peripheral action (Edwards 1977).

Regulation of Sodium Excretion

The renal excretion of sodium is controlled and influenced by many factors, some arising outside the kidney, and some direct renal responses to changes in ECF volume (Reviewed Schrier and Anderson 1980, Findlay 1986).

Renal Factors: Whilst alterations in filtered sodium load play a part in determining sodium excretion, they probably do not have a major role. Intrarenal haemodynamic factors are undoubtedly important, however. An increase in renal perfusion pressure, as during expansion of the ECF volume, and haemodilution with a fall in plasma oncotic pressure, both increase the hydrostatic pressure around the proximal and distal nephron. This decreases sodium reabsorption and causes a natriuresis. Thus an increased urinary sodium excretion during administration of hypertonic saline, for example, may be a normal homeostatic response and should not be interpreted as a "salt-wasting" disorder. Intrarenal vasodilatation causes natriuresis by the same mechanism. Bradykinin and prostaglandin E_2 are both synthesised in the kidney and are potent renal vasodilators. Their importance in natriuretic conditions is uncertain.

Extrarenal Factors: The kidneys are richly innervated by the adrenergic system via the renal nerves, which decrease sodium excretion by several mechanisms. They are stimulated by input from high and low pressure thoracic sensors, and are probably the most important means by which the brain modulates sodium excretion. Adrenergic stimulation also influences sodium excretion by changes in central haemodynamics. Adrenergic activity is increased in stressed neurosurgical patients (Neil-Dwyer *et al.* 1974) and will contribute to changes in salt and water balance. *The renin-angiotensin-aldosterone system* has synergistic vasoconstrictor and sodium-retaining actions, and has a central role in the renal regulation of sodium balance. It normally protects against contraction of the ECF volume. The system is activated by release of renin from the juxtaglomerular apparatus in the kidneys. This is stimulated by a fall in renal perfusion pressure, sensed by baroreceptors in the renal afferent arteries, by the sympathetic nervous system through the renal nerves, and by a fall in the sodium supply to the distal renal tubule (Reviewed: Mulrow 1976, Schrier and Anderson 1980, Findlay 1986).

Circulating Natriuretic Factors: Several groups reported that plasma and urine from volume-expanded animals and man inhibited net renal tubular sodium transport, and caused a natriuresis. De Wardener *et al.* (de Wardener 1982) demonstrated the presence of a renal Na^+-K^+ ATPase inhibitor in human plasma, whose concentration was related to sodium intake. A digoxin-like substance has been proposed, which may originate in the hypothalamus. Chemical properties of the plasma extracts, however, suggested that the agent might be a peptide (de Wardener and Clarkson 1982). It seems that there may be more than one circulating natriuretic factor. An exciting recent discovery has been the identification of α-human atrial natriuretic peptide (α-h ANP). This has 28 amino acids and the cardiac atria are the major sites of production. Small amounts are also synthesised in the brain and kidneys. Plasma ANP levels in man change rapidly in response to acute changes in sodium balance, and an increased concentration preceded the natriuresis induced by saline infusion (Leaders, Lancet 1984 and 1986, Singer 1987). When synthetic α-h ANP was infused in man to achieve similar plasma concentrations, a striking increase in urinary sodium and water excretion resulted, and on a molar basis α-h ANP was 1,000 times more potent than frusemide (Anderson 1986). It is proposed that α-h ANP is released into the blood by stretching of the atria after expansion of the ECF volume. It then acts on the kidney to promote excretion of salt and water, and thereby restores homeostasis. The mechanism underlying natriuresis is uncertain. The peptide does not inhibit Na^+-K^+ ATPase, and its relation to de Wardener's less-well characterized natriuretic factor is unknown.

Changes of Sodium and Water Balance Following Surgery

Water and sodium retention are a normal occurrence during the first 24–36 hours after surgery, and on the basis of careful fluid balance studies in the 1950's (le Quesne and Lewis 1953) were attributed to increased secretion of antidiuretic hormone and adrenocorticosteroids. Using a radioimmunoassay for plasma AVP, Haas and Glick (1978) showed that in patients undergoing general surgery, AVP was released during operation, with plasma concentrations ranging from 12.5–82 pg/ml. The levels fell gradually during the first 3 or 4 days post-operatively. The increase was unrelated to plasma osmolality. Many factors probably contribute to AVP release peri-operatively. Circulatory changes induced by haemorrhage, hypotensive agents, and tilting (Aziz and Forsling 1979, Cochrane *et al.* 1981) are probably important, and in some patients, nausea or emesis, or mannitol infusion contribute. It is important to be aware of the increased AVP secretion which occurs in response to trauma, and to infuse hypotonic fluids cautiously after surgery.

Injudicious intravenous infusion, and not inappropriate ADH secretion, probably accounts for many cases of hyponatraemia in neurosurgical patients (Bouzarth and Shenkin 1982). Volume expansion and not "cerebral natriuresis" would account for the increased urinary sodium excretion frequently observed.

Fluid Balance Disturbances in Neurosurgical Patients Caused by Damage to the Neuroendocrine System

Some working definitions for the more commonly encountered disorders are presented (Table 1), and relevant reference ranges (Appendix 1). Meticulous fluid intake and output records are essential for early detection and management of fluid disturbances. Daily weighing is helpful if it is accurate, but this is seldom the case in the normal ward setting. Frequent estimations of plasma sodium are necessary, and paired serum and urinary osmolality measurements if there is evidence of a fluid disturbance. Unless the patient's ECF volume status is known, 24-hour urinary sodium excretion can be very difficult to interpret. Attempts to replace urinary sodium losses with infused saline can be hazardous in a patient who is having a natriuresis

Table 1. *Definitions**

Polyuria:
  More than 3 l urine/24 h (adults)

Complete cranial diabetes insipidus:
  Lack of circulating AVP
  Polyuria approximately 10% of filtered water, adults: generally
    >10–12 l/24 h
  Urinary osmolality below 100 mosm/kg water
  Plasma sodium and osmolality are raised if fluid intake is inadequate.

Partial cranial diabetes insipidus:
  Impaired secretion of AVP
  Variable polyuria, adults: 3–6 l/24 h
  Urine hypoosmolar with respect to plasma. *BUT* during fluid
    deprivation, urinary osmolality may exceed 300 mosmol/kg water
  Plasma AVP: In five studies within the range <0.8 to 1.3 pg/ml
    (Edwards, 1979).

Inappropriate secretion of antidiuretic hormone (SIADH):
  Secretion of AVP despite a low plasma osmolality.
  Expansion of the ECF volume
  No discernable stimulus, such as dehydration, hypovolaemia, hypotension, drugs
  Normal adrenal, renal, and thyroid function

Plasma:
  Sodium low
  Osmolality low
  AVP detectable *when hyponatraemic*

Urine:
  Osmolality inappropriately high
  Sodium greater than 20 mmol/l

---

  * Schrier and Leaf 1981, Andreoli 1982.

because he is volume expanded. Calculation of the free water clearance provides a useful indicator of antidiuretic activity (Whitaker *et al.* 1985).

Diabetes Insipidus (DI): Severe DI is readily diagnosed. However, many patients with cranial DI have a partial defect in AVP secretion. These patients can often concentrate their urine to some extent during fluid deprivation. It can be difficult at times to differentiate between partial DI and primary polydipsia, which may occur in patients with hypothalamic pathology. Chronic excessive water intake diminishes the renal medullary concentration gradient upon which the antidiuretic action of AVP depends, and patients with polydipsia are unable to concentrate their urine maximally. Measurement of plasma sodium and osmolality is helpful sometimes, since in DI the values may be towards or a little above the upper limit of normal, whereas in polydipsia values are usually near the lower end of the range (Zerbe *et al.* 1981). An indirect test is

to compare the maximum urinary osmolality achieved during dehydration to that achieved after vasopressin injection (as desamino-D-arginine vasopressin, DDAVP) (Appendix 2). In theory, patients with DI concentrate the urine better after vasopressin than after dehydration, whereas patients with primary polydipsia concentrate similarly with both. However, Zerbe *et al.* found that the test was sometimes an unreliable discriminant. A therapeutic trial of DDAVP may help in the differentiation, but this can induce severe water intoxicaton in patients with polydipsia. The most reliable discrimination can be achieved by measuring plasma AVP during infusion of hypertonic saline over several hours. However, there is a risk of inducing fluid overload, and the infusion itself may cause a solute diuresis making interpretation of urinary flow changes difficult (Zerbe *et al.* 1981). Very few laboratories could return AVP results sufficiently quickly to be of value in most postoperative situations in which the diagnostic problem arises. DI is frequently transient and short-lived in neurosurgical patients. It is a common occurrence following transsphenoidal adenectomy (Whitaker *et al.* 1985), and probably reflects the extreme sensitivity of the hypothalamic-neurohypophyseal unit to local alterations in blood flow, oedema, and traction on the pituitary stalk (Cobb 1980).

The syndrome of inappropriate ADH secretion (SIADH) was a term used originally in 1957 to describe the effects of ectopic production of an antidiuretic substance by bronchial carcinoma (Schwartz *et al.* 1957). It was subsequently reported to occur in malfunctions of the nervous system (Goldberg and Handler 1960, Lester and Nelson 1981). However, the condition has been over-diagnosed in patients with neurological disorders, and has been applied indiscriminately to patients with hyponatraemia. Strict criteria should be fulfilled (Table 1), and the diagnosis should be considered only after other causes of hyponatraemia have been excluded, and in the absence of hypovolaemia, oedema, adrenal insufficiency, hypothyroidism, renal failure, and drugs which impair water excretion. When these criteria have been met, AVP levels have generally been inappropriately high for the hypotonic plasma. An unambiguous diagnosis of SIADH can only be made by measuring AVP while the patient is hyponatraemic. Results from samples obtained after correcting the salt and water balance are difficult to interpret. Several different mechanisms appear to underlie SIADH in patients with brain disorders.

A triphasic pattern of development of DI has been described in animals and man following hypothalamic damage or high pituitary stalk section with more than ninety per cent interruption of axoplasmic flow in the median eminence and below (Randall *et al.* 1960, Cobb 1980). Transient DI lasting 1–7 days is followed by an "interphase" in which the urinary loss decreases and there may be inappropriate water retention (SIADH). This is postulated to result from leakage of AVP from degenerating pituitary cells. Permanent DI follows within a few days. AVP leakage from disintegrating cells was proposed as an explanation for severe hyponatraemia which developed in three patients following transsphenoidal pituitary adenectomy (Cusick *et al.* 1984). Our own group found no evidence for this in 35 patients who had pituitary adenectomy (Whitaker *et al.* 1985).

Hyponatraemia in Hypothyroidism and Adrenal Insufficiency

Recent studies in hypothyroid man and rats with impaired water excretion have demonstrated an increase in immunoreactive AVP, indicating some disturbance in hypothalamic-neurohypophyseal function. In addition, decreased proximal tubular reabsorption may increase urinary sodium loss in this condition (Schrier 1980). The pathophysiology of hyponatraemia in adrenal insufficiency is imperfectly understood. In primary adrenal insufficiency, a deficiency of mineallocorticoids leads to excessive sodium loss, a fall in the ECF volume, and an increase in AVP secretion, as well as to diminished renal perfusion pressure and altered renal haemodynamics. These factors together would account for hyponatraemia. However, glucocorticoid deficiency alone also leads to impaired water excretion, in the absence of volume depletion. AVP levels were found to be inappropriately increased in glucocorticoid deficiency (Boykin *et al.* 1978). It was suggested that impaired cardiac function might lead to baroreceptor-mediated stimulation of AVP release. It has also been proposed that glucocorticoids are important in allowing maximal suppression of AVP release, as well as maximal impermeability of the collecting ducts. However, most studies used pharmacological doses of glucocorticoids, and the physiological significance of the observations is uncertain (Kleeman *et al.* 1960). Four of 35 patients studied in Southampton after pituitary adenectomy had evidence of glucocorticoid insufficiency during the first week following surgery, despite receiving steroid cover. Three became hyponatraemic, and one had an acute hypotensive episode (Whitaker *et al.* 1985). The steroid regimen was subsequently mod-

ified so that the steroid doses were spaced to provide better 24-hour cover. Steroid withdrawal must be undertaken very carefully in patients who have had adenectomy for Cushing's disease, because they have prolonged suppression of their pituitary-adrenal axis which may persist for over a year after surgery (Fitzgerald *et al.* 1982, Boggan *et al.* 1983).

Many factors may contribute to fluid balance disturbance in neurosurgical patients. These all have to be considered when planning appropriate corrective therapy.

References

Anderson J, Struthers A, Christofides N, Bloom S (1986) Atrial natriuretic peptide: an endogenous factor enhancing sodium excretion in man. Clinical Science 70: 327–331

Andreoli TE (1982) Posterior pituitary: antidiuretic hormone. In: Wyngaarden JB, Smith LH (eds) Cecil textbook of medicine, 16th Ed. WB Saunders, Philadelphia, pp 1192–1198

Aziz LA, Forsling ML (1979) Anaesthesia and vasopressin release in the rat. J Endocrinol 81: 123 p

Boggan JE, Tyrrell JB, Wilson CB (1983) Transsphenoidal microsurgical management of Cushing's Disease. Report of 100 cases. J Neurosurg 59: 195–200

Bouzarth WF, Shenkin HA (1982) Is cerebral hyponatraemia iatrogenic? Lancet 1: 1061–1062

Boykin J, de Torrenté A, Erickson A, Robertson G, Schrier RW (1978) Role of plasma vasopressin in impaired water excretion of glucocorticoid deficiency. J Clin Invest 62: 738–744

Cobb WE (1980) Endocrine management after pituitary surgery. In: Post KD, Jackson IMD, Reichlin S (eds) The pituitary adenoma. Plenum Press, New York, pp 417–435

Cochrane JPS, Forsling ML, Gow NM (1981) Arginine vasopressin release following surgical operations. Br J Surg 68: 209–213

Cort JH, Yale MD (1954) Cerebral salt wasting. Lancet i: 752–754

Cusick JF, Hagen TC, Findling JW (1984) Inappropriate secretion of antidiuretic hormone after transsphenoidal surgery for pituitary tumors. N Eng J Med 311: 36–38

De Wardener HW (1982) The natriuretic hormone. Ann Clin Biochem 19: 137–140

— Clarkson EM (1982) The natriuretic hormone: recent developments. Clin Sci 63: 415–420

Edwards CRW (1977) Vasopressin. In: Martin L, Besser GM (eds) Clinical neuroendocrinology. Academic Press, New York, pp 527–567

— (1979) Vasopressin: In: Gray CH, James VHT (eds) Hormones in blood, 3rd Ed, Vol 2. Academic Press, London, pp 423–450

Findlay ALR (1986) Sodium excretion. In: Lightman SL, Everitt BJ (eds) Neuroendocrinology. Blackwell Scientific Publications, Oxford, pp 229–251

Fitzgerald PA, Aron DC, Findling JW, Brooks RM, Wilson CB, Forsham PH, Tyrrell JB (1982) Cushing's Disease: transient secondary adrenal insufficiency after selective removal of pituitary microadenomas; evidence for a pituitary origin. J Clin Endocrinol Metab 54: 413–422

Goldberg M, Handler JS (1960) Hyponatraemia and renal wasting of sodium in patients with malfunction of the central nervous system. N Eng J Med 263: 1037–1043

Haas M, Glick SM (1978) Radioimmunoassayable plasma vaso-pressin associated with surgery. Arch Surg 113: 597–600

Jenkins JS, Mather HM, Ang V (1980) Vasopressin in human cer-ebrospinal fluid. J Clin Endocrinol Metab 50: 364–367

Kleeman CR, Koplowitz J, Maxwell MH, Cutler R, Dowling JT (1960) Mechanisms of impaired water excretion in adrenal and pituitary insufficiency. II Interrelationships of adrenal cortical steroids and antidiuretic hormone in normal subjects and in diabetes insipidus. J Clin Invest 39: 1472–1480

Le Quesne LP, Lewis AAG (1953) Postoperative water and sodium retention. Lancet i: 153–158

Leader (1984) Atrial natriuretic peptides. Lancet ii: 328–329

— (1986) Atrial natriuretic peptide. Lancet ii: 371–372

Lester MC, Nelson PB (1981) Neurological aspects of vasopressin release and the syndrome of inappropriate secretion of antidi-uretic hormone. Neurosurgery 8: 735–740

Lightman SL, Everitt BJ (1986) Water excretion. In: Lightman SL, Everitt BJ (eds) Neuroendocrinology. Blackwell Scientific Publications, Oxford, pp 197–206

Morton JJ, Padfield PL, Forsling ML (1975) A radioimmunoassay for plasma arginine vasopressin in man and dog: application to physiological and pathological states. J Endocrinol 65: 411–424

Mulrow PJ (1976) Renal hormones. In: Brenner BM, Rector FC Jr (eds) The Kidney, Vol 1. WB Saunders, Philadelphia, pp 477–492

Neil-Dwyer G, Cruikshank J, Stott A, Brice J (1974) The urinary catecholamine and plasma cortisol levels in patients with sub-arachnoid haemorrhage. J Neurol Sci 22: 375–382

Nelson PB, Seif SM, Maroon JC, Robinson AG (1981) Hypona-traemia in intracranial disease: perhaps not the syndrome of inappropriate secretion of antidiuretic hormone (SIADH). J Neurosurg 55: 938–941

Randall RV, Clark EC, Dodge HW, Grafton Love J (1960) Polyuria after operation for tumors in the region of the hypophysis and hypothalamus. J Clin Endocrinol 120: 1614–1621

Rap ZM, Chwalbinska-Moneta J (1978) Vasopressin concentration in blood during acute short-term intracranial hypertension in cats. Adv Neurol 20: 381–388

Schrier RW (1980) Disorders of water metabolism. In: Schrier RW (ed) Renal and electrolyte disorders, 2nd Ed. Little, Brown, and Company, USA, pp 1–64

— Anderson RJ (1980) Renal sodium excretion, edematous disor-ders, and diuretic use. In: Schrier RW (ed) Renal and electrolyte disorders, 2nd Ed. Little, Brown, and Company, USA pp 65–114

— Leaf A (1981) Effect of hormones on water, sodium, chloride, and potassium metabolism. In: Williams RH (ed) Textbook of endocrinology, 6th Ed. WB Saunders, Philadelphia, pp 1032–1046

Schwartz WB, Bennett W, Curelop S, Bartter FC (1957) A syndrome of renal sodium loss and hyponatremia probably resulting from inappropriate secretion of antidiuretic hormone. Am J Med 23: 529–542

Singer DRJ (1987) Atrial natriuretic peptides: clues to their physi-ological and clinical importance. Postgrad Med J 63: 1–4

Whitaker SJ, Meanock CI, Turner GF, Smythe PJ, Pickard JP, Noble AR, Walker V (1985) Fluid balance and secretion of antidiuretic hormone following transsphenoidal pituitary surgery. J Neu-rosurg 63: 404–412

Zerbe RL, Baylis PH, Robertson GL (1981) Vasopressin function in clinical disorders of water balance. In: Beardwell C, Robertson GL (eds) Butterworth international medical reviews, clinical en-docrinology 1. The pituitary. Butterworth and Co (Publishers) Ltd, London, pp 297–329

Correspondence: V. Walker, M.D., Chemical Pathology and Human Metabolism, Southampton University Medical School, Southampton, SO9 4XY, England.

Appendix 1
Biochemical reference ranges for young adults

Plasma	sodium	135–145 mmol/l (135–145 mEq/l)
	chloride	95–105 mmol/l (95–105 mEq/l)
	osmolality	280–295 mosmol/kg water
	urea	3.0–6.5 mmol/l (18–40 mg/100 ml)
	creatinine	60–125 µmol/l (0.68–1.41 mg/100 ml)
Urine	sodium	: varies with intake, generally around 100–200 mmol/24 h
		: a concentration of < 10 mmol/l suggests hypovolaemia or sodium depletion
	osmolality	: varies with fluid intake
		: after 16–18 hours of fluid deprivation, most healthy adults concentrate urine to around 1,000 to 1,200 mosmol/kg water
		: values around 290 mosmol/kg are isotonic with plasma.
	urea	: 250–500 mmol/l (1.5–3.0 g/100 ml).
	creatinine	: 9.0 to 18.0 mmol/24 h (1–2 g/24 h).

AVP (arginine vasopressin; the human antidiuretic hormone)

*Plasma** (Morton *et al.* 1975; Edwards 1979)	: half life around 7–8 min
	: partially hydrated subjects, 1–5 pg/ml (0.9–4.6 pmol/l).
	: water-loaded, around 0.5 pg/ml (0.5 pmol/l) or less
	: overnight fluid deprivation (12 h), 7.4 ± 1.5 pg/ml.
	: 24 hour fluid deprivaton, 8.8 ± 0.6 pg/ml.
Lumbar CSF (Jenkins *et al.* 1980)	: 2.4 ± 0.7 pg/ml

* Concentrations vary for different laboratories. Some representative published ranges are presented.

Note: 1 pg/ml AVP = 0.92 pmol/l, 1 µU of AVP = 2.5 pg = 2.3 × 10⁻³ pmol.

Appendix 2
Combined water deprivation/DDAVP (desmopressin) test

Preparation: Allow only *water* to drink for the 8 hours preceding the test since tea or coffee can produce spurious results. No smoking is permitted. Record the patient's weight at the beginning of the test, and terminate the test if weight loss exceeds 3 per cent.

The patient may eat and drink after the DDAVP injection, but to avoid water intoxication, the hourly fluid intake should not exceed the urinary volume passed in the preceding hour.

The test: It is safest to do the test during the daytime, starting, for example, at 6.00 or 8.00 h.

6.00 h.   Weigh. Collect urine and blood.

6.00 h to 14.00 h. Deprive of all fluids for 8 hours. Weigh, and collect urine and blood at two-hourly intervals.

14.00 h.   Give 2 µg DDAVP intramuscularly or 40 µg intranasally. Collect urine and blood at *2 hours* and *4 hours* after DDAVP. Urine and plasma osmolalities are measured.

Interpretation

Normals: Concentrate urine to more than 700 mosmol/kg water after 8 hours of fluid deprivation, and there is little further increase with DDAVP. Plasma osmolality remains normal.

An osmolality of < 500 mosmol/kg water is abnormal.

Values between 500 and 700 are equivocal.

If DDAVP produces an increase in urinary concentration of over 10 per cent, cranial diabetes insipidus is likely. The serum osmolality may be > 295 mosmol/kg water in DI.

If DDAVP produces only a small increase in urinary concentration (< 5 per cent), primary polydipsia is likely.

Acta Neurochirurgica, Suppl. 47, 102–110 (1990)

The Stress Response in Subarachnoid Haemorrhage and Head Injury

G. Neil-Dwyer, J. M. Cruickshank, and **R. Doshi**

Department of Neurosurgery, Brook General Hospital, London, U.K.

Introduction

The concept of the stress response—the response to a variety of non-specific damaging agents—has been known for years. There is evidence in some older medical texts that physicians had long suspected the beneficial effect of this response. Hippocrates made mention of this in his writings (Jones 1923) when he stated that "disease caused nature the constitution of the individual to make every effort she could through an exciting cause to restore the original status". John Hunter in 1794 was the first to put forward the idea of the response to injury as a concept when he wrote "there is a circumstance attending accidental injury which is not belonged to disease namely that the injury alone has in all cases a tendency to produce both the disposition and the means of cure". Subsequently, this response was described by many eminent workers and it was left to Hans Selye in 1936 who, while working with animals, showed that when they were exposed to a variety of non-specific damaging agents they responded with a discharge of adrenaline and adrenal cortical hormones. The most striking feature of the response was its non-specific nature. Magoun, Ranson, and Hetherington (1937) at around this time, found that stimulating the hypothalamus increased the production of adrenaline and noradrenaline. Their results pointed to the hypothalamus as being involved in the so called "adaptive reaction". However, in 1944 it became apparent to Selye and, subsequently in 1966 to Raab in particular, that under certain circumstances increased endogenous production during stress or exogenous administration of ACTH, corticosteroids and catecholamines can, in their own right, become the cause of disease, that is, hypertension, diabetes, myocarditis etc. These diseases were well described by Selye

as "the diseases of adaption". Further progress with this concept occurred when Oka (1956) in 1956 was able to show that a marked increase in the free 17 hydroxycorticosteroids after cerebro-vascular accidents indicated a poor prognosis and Kerr Corbett, Prys-Roberts *et al.* in 1968 demonstrated that overactivity of the sympathetic nervous system in tetanus was a factor in the morbidity of that disease.

Subarachnoid haemorrhage and head injury are acute illnesses. Some of the systemic effects, particularly in subarachnoid haemorrhage, such as hypertension, electrolytic changes, proteinuria, glycusuria and electrocardiographic (ECG) abnormalities had been recognized for years but the mechanism causing these changes had been a matter for speculation.

Subarachnoid Haemorrhage

ECG Abnormalities

In considering the ECG abnormalities, most experimental evidence pointed towards the possible action of the sympathetic nervous system associated with abnormal hypothalamic activity but direct evidence was lacking. The hypothesis that we formulated was that the overactivity of the adrenal gland and the sympathetic nervous system was an integral part of a subarachnoid haemorrhage and that this overactivity by acting on the heart, kidney and brain was a factor in the morbidity and mortality of this condition. The purpose of a series of studies over a number of years was to test this hypothesis.

Some years ago we looked at a group of 40 patients who had sustained a spontaneous subarachnoid haemorrhage. There were 16 men (mean age 41) and 24 women (mean age 42). The patients were studied over

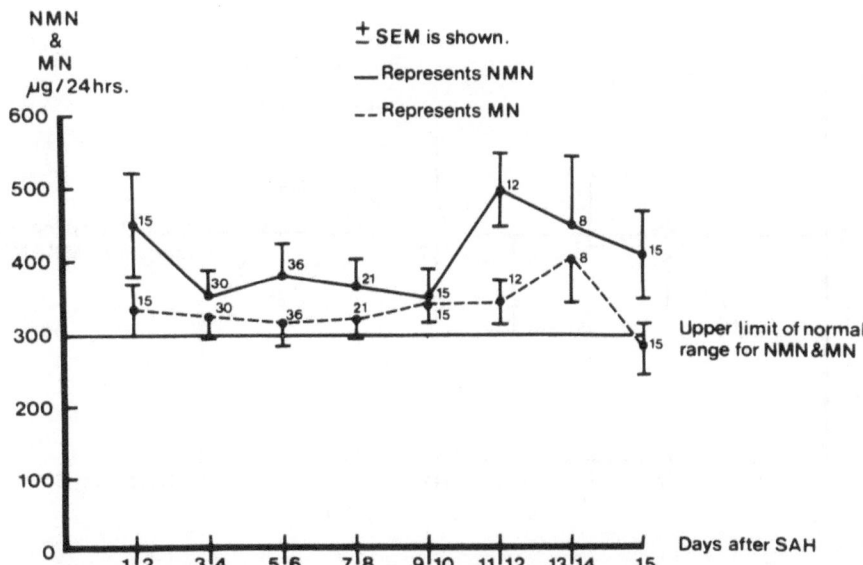

Fig. 1. Values of the serial urinary norme-tanephrine and metanephrine levels over 15 days after SAH

the period of their stay in hospital, on average two weeks. Certain serial investigations were done over the course of this time. A daily clinical assessment, blood pressure and pulse rate recordings, a 24 hour urine collection was started for the estimation of normeta-nephrine, the breakdown product of noradrenaline and metanephrine, the breakdown product of adrenaline, and blood was also taken for plasma cortisol estima-tions. Figure 1 shows the values of the serial urinary normetanephrine and metanephrine levels during the period of the study following the haemorrhage and demonstrates that on no occasion did the mean values of normetanephrine and metanephrine lie within the normal range. A similar picture occurred with the serial measurements of plasma cortisol.

A total of 197 ECGs were recorded, an average of

5 per patient, and the types of ECG abnormalities noted and their incidence is shown in Table 1. For analytical purposes, the patients were divided into thre groups. Patients with all normal ECGs [8], those with all ab-normal ECGs [19] and those with changing ECGs [13]. The relationship between the urinary catecho-lamines and plasma cortisol levels with time in the normal and abnormal ECG groups is shown in Fig. 2 (a and b) and 3 (a and b) respectively. It is seen that in the normal ECG group, though the number of meas-urements are relatively small, catecholamine and plasma cortisol levels are within or marginally above the normal range throughout the period of study. This is in distinct contrast to the abnormal ECG group of patients were the catecholamine and plasma cortisol levels remained above the normal range.

The ECG features were broken down into com-ponents in an attempt to relate high urinary catecho-lamine levels to particular types of abnormality. ECG components such as short PR intervals, large T waves, long QTC intervals and large U waves were found to be associated with significantly higher levels of urinary catecholamines.

Clearly, a case for catecholamines being responsible for the ECG abnormalities in subarachnoid haemor-rhage could be made though a cause and effect rela-tionship is not proven.

The mode of action of the catecholamines upon the heart is not fully understood. Both transient and more permanent effects can occur. Transient effects stem from the properties of the catecholamines and, if car-ried to excess, produce pathological changes such as

Table 1. *Types of ECG abnormality and their incidence in 40 patients*

Type of abnormality	No. of patients	%
Tachycardia (100/min)	13	33
Bradycardia (60/min)	9	23
P pulmonale	16	41
Short P-R (0.13 sec)	10	25
Q waves	2	5
Raised ST segment	13	33
Depressed ST segment	3	8
Coved S-T	6	16
Inverted T	14	36
Peaked T	8	20
Tall U (0.1 mv)	13	33
Long Q-Tc (0.43 sec)	22	55

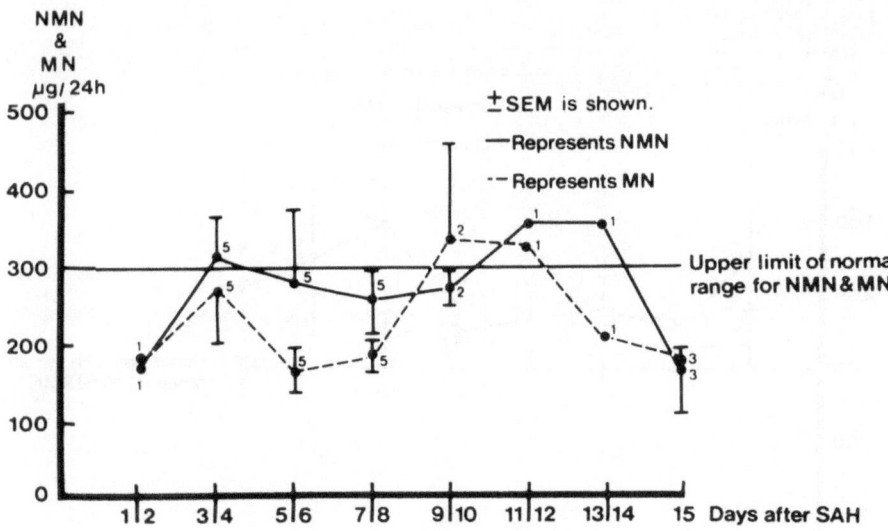

Fig. 2 a. Urinary catecholamine levels in normal ECG patients

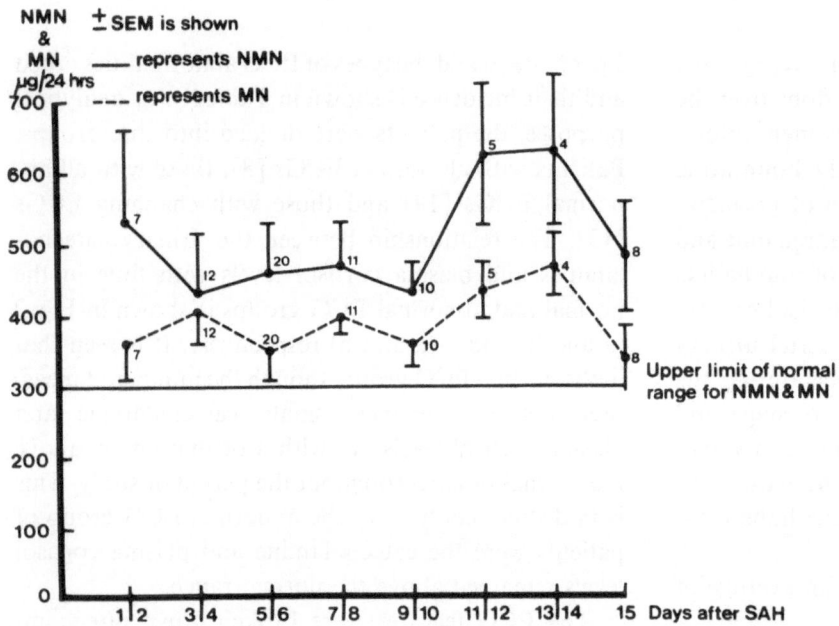

Fig. 2 b. Urinary catecholamine levels in abnormal ECG patients

myocardial necrotic lesions. Cardiac myofibrillar degeneration, frequently focal and subendocardial, has been well described in association with pheochromocytomas and catecholamine administration.

Renin/Angiotensin II System

A part of the initial hypothesis had been that the kidneys could be affected by the increased sympathetic activity produced by this condition. Stimulation of the limbic system in animals had been shown to produce vasomotor, cellular and functional changes in the kidney (Hoff, Kell, and Hastings 1951). Stimulation of

the posterior aspect of the hypothalamus had been shown to produce renal vasoconstriction (Takeuchi, Yagi, Nakayama *et al.* 1960) and, in addition, hypothalamic stimulation had increased renin output from the kidney (Zanchetti 1976). James, in 1972, demonstrated that renal blood flow was significantly decreased following a subarachnoid haemorrhage despite systemic hypertension and had argued that it was due to renal arterial vasoconstriction. It was, therefore, possible that increased activity of the renin/angiotensin II system in subaracharchnoid haemorrhage may, by its inter-relationship with the sympathetic nervous system, be an added factor in the morbidity and mortality of

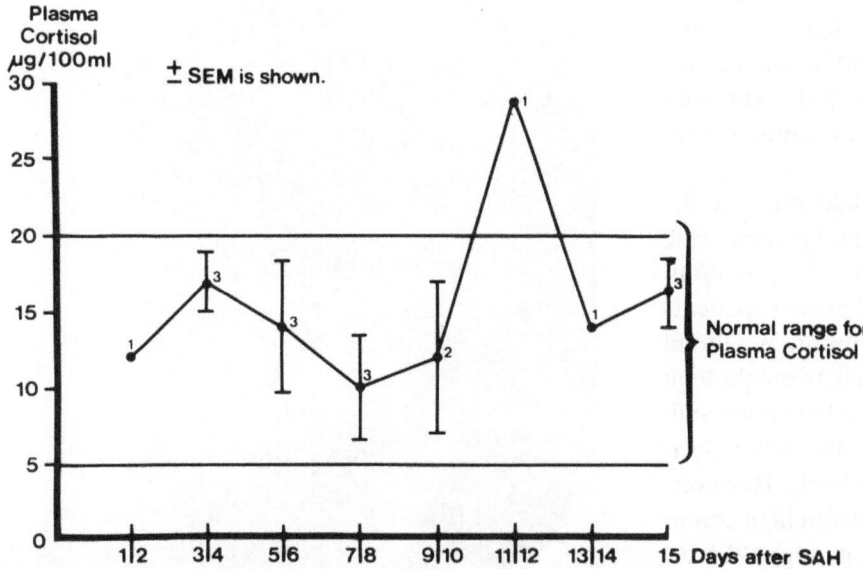

Fig. 3 a. Plasma cortisol levels in normal ECG patients

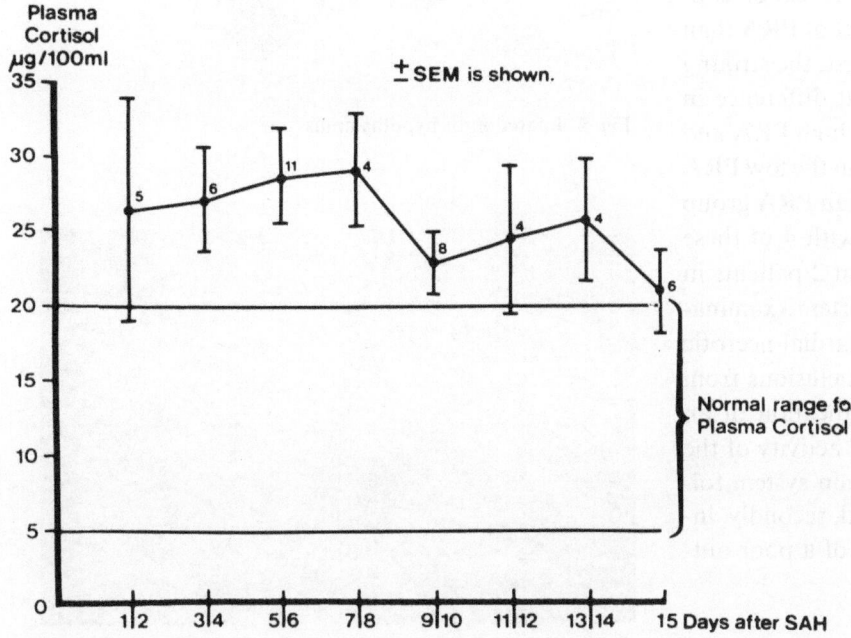

Fig. 3 b. Plasma cortisol levels in abnormal ECG patients

this condition. We decided to examine the possibility. 6 post-laminectomy patients (age range 24–63 years) were used as controls in order to obtain a normal range of plasma renin activity (PRA) for patients on strict bed rest. There were 4 women. 23 patients (age range 24–64 years) who had suffered a subarachnoid hae-morrhage, 17 of these proving to have aneurysms, were studied. There were 15 women. All patients were on strict bed rest throughout the period of study. Controls and SAH patients were age matched as far as possible to take account of the observation that PRA declines with age (Crane and Harris 1976). Patients who had

previously been on hypotensive or diuretic medication were excluded. Both control and SAH patients were put on a diet containing 120 mEq sodium and 40–50 mEq potassium daily.

All patients were on strict bed-rest throughout the period of study. On the first, fourth and eight days between 8.00 and 9.30 hrs the patients having been fasted over night, their clinical state was assesed, the blood pressure measured and blood was taken for the estimation of plasma renin activity, urea and serum electrolytes. A 24-hour urine collection was started to measure the excretion rates of urinary catecholamines—namely adrenaline, noradrenaline, 3-methoxy, 4-hydroxymandelic acid (HMMA) metadrenaline and urinary sodium. Post mortem examination was carried out on all patients who died, with particular attention being paid to the histology of the hypothalamus, kidney and heart.

Results showed that 13 patients had what we describe as low PRA (<1.77 p mol/ml/hr), 2 patients were within the normal range (>1.77 and <2.5 p mol/ml/ hr) and 8 patients had high PRA and these were patients whose renin levels were >2.5 p mol/ml/hr. We found that there was a significant correlation between the high PRA and the high levels of urinary catecholamines and, curiously, the low PRA was also significantly associated with high urinary noradrenaline levels. However, there was a significant correlation between high adrenaline, HMMA, metadrenaline levels and high PRA estimations. Although the SAH patients with neurological deficits and those with an altered levels of consciousness had a significantly higher level of PRA than did patients without a neurological deficit, the striking feature in this study was the significant difference in the outcome between the patients in the high PRA and the low PRA groups. Of the 13 patients in the low PRA group, 11 did well, 2 died, while in the high PRA group only 2 patients did well, 6 did poorly with 4 of these dying. An interesting feature was that in 2 patients in the high PRA group who had post mortem examination, both had hypothalamic and myocardial necrotic lesions. We felt that there were two conclusions from this study, firstly that there was an indication of an inter-relationship between the increased activity of the sympathetic nervous system and the renin system following a subarachnoid haemorrhage and, secondly, increased PRA appears to be a predictor of a poor outcome after subarachnoid haemorrhage.

Hypothalamic and Myocardial Lesions

One of the feasible explanations of this increased sympathetic drive is an abnormal hypothalamic response induced by spasm of the small vessels supplying the hypothalamus after a subarachnoid haemorrhage. Therefore, we looked at the hypothalamus and myocardium in 54 patients who had died following a subarachnoid haemorrhage. Lesions in the hypothalamus consisted of small perivascular haemorrhages, distensions of perforated vessels with small ball haemorrhages, oedema of the vessel walls involving the endothelial cells with perivascular cuffing of polymorpholeucocytes, microinfarction and, extraordinarily, in

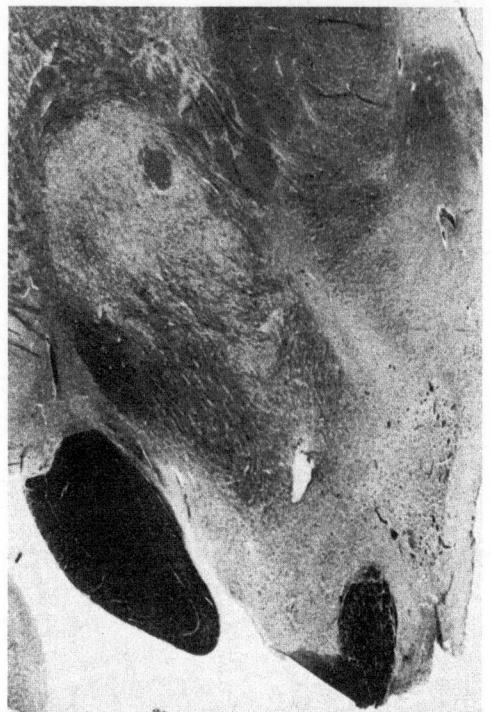

Fig. 4. Infarction of hypothalamus

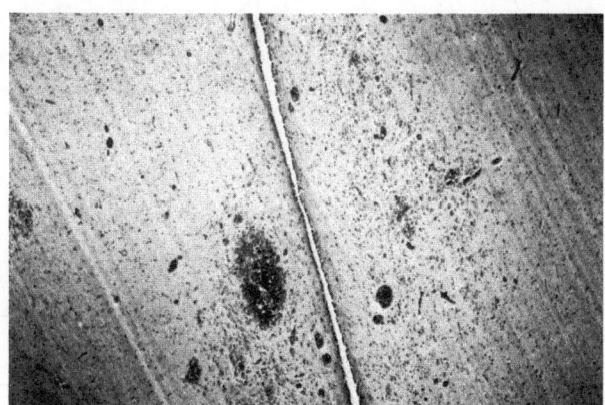

Fig. 5. Hypothalamic haemorrhages

three cases complete infarction of the hypothalamus (Fig. 4 and 5). The histology of the majority of the myocardial lesions found in the left ventricle were foci of necrotic muscle fibres (Fig. 6). The age range of the 54 patients was 12–73 years and, interestingly, hypothalamic and extensive myocardial lesions were found in the youngest patient, a boy of twelve. Table 2 shows the relationship between the hypothalamus and myocardial lesions, 42 of the 54 patients having this combination.

Fig. 6. Foci of necrotic myocardial muscle fibres

Table 2. *Relationship between the hypothalamic and myocardial lesions*

	Total	Male	Female
Hypothalamic and myocardial lesions	42	16	26
Hypothalamic lesions only	7	—	7
Neither lesion	5	5	—

Reversal of ECG Abnormalities and Myocardial Lesions

If the hypothesis that the ECG changes occurring in subarachnoid haemorrhage are due to increased sympathetic activities was correct, then it should be possible to reverse them. As most of the ECG changes probably result from the beta-adrenergic activity of the catecholamines, the effect of the beta-blocker propranolol on the changes was investigated. 80 mg of oral propranolol was given to 6 cases of spontaneous subarachnoid haemorrhage who had at least one of the previously mentioned catecholamine-linked ECG abnormalities. Table 3 shows the overall results. The P-wave peaking

Table 3. *Effect of propanolol on ECG changes in 6 patients*

ECG abnormalities	Patients	Results
Peaked P-wave	3	all abolished
Short P-R interval	3	all lengthened
Tall V-wave	2	both abolished
Long Q-Tc	5	all shortened
S-T Segment	3	variable
Path Q-wave	1	not affected
Inverted T-wave	1	not affected
Peaked T-wave	1	increased, abolished by atropine

seen in 3 of the six cases was abolished. The short PR intervals in 3 cases was lengthened to normal, the tall U waves seen in 2 cases were diminished in amplitude while the long QTC interval was shortened in all 5 cases in which it occurred. Interestingly, the peaking of the T wave was made worse by propranolol and this was reduced to a normal amplitude when the patient was given atropine. This indicates that it is not only increased sympathetic action but clearly there is some altered parasympathetic activity as well.

The question then was, having altered the ECG changes was it possible to alter the histological abnormalities found in the heart? In a recent randomized double-blind between patients' study in which there were 104 patients, 18 died. Of these 18, 16 had post mortems. There were 9 patients in the control group who had received placebo tablets and 7 in the treated group who had received both propranolol and phentolamine, an alpha-blocker. Table 4 shows the results. We felt that the absence of myocardial necrotic lesions in the treated group was due to propranolol rather than the phentolamine because of the relative absence of alpha-receptors in the myocardium.

Head Injury

Head injuried patients exhibit many of the systemic effects noted following a subarachnoid haemorrhage.

Table 4. *Post-mortem results*

	β-blocker patients (7)	Placebo patients (9)
Coronary arteries	2—no atheroma 5—mild atheroma	2—no atheroma 7—mild atheroma
Left ventricle	no lesions	1—no lesions 2—focal necrotic lesions of muscle fibres 6—as above and inflammatory cell infiltrations
Hypothalamus	7 lesions	9 lesions

Patients with severe head injury are known to be markedly hyper adrenergic and can experience cardiac morbidity. In a small study of 7 patients, all male, mean age 22, with severe diffuse brain injury admitted into the Neurosurgical Unit within 4 hours from the time of accident, each patient was studied intensively for 72 hours. Continuous recording of the electrocardiagram was made using an Oxford-medilog recorder and analysed by using an Oxford analysis system and a Reynolds Pathfinder. Daily assays of creatine kinase and its myocardial isoenzymes CKMB were made and 24-hour urine samples were collected for assay of free catecholamines and their metabolites. This study showed increased levels of urinary catecholamines, grossly high levels of total CK and significantly raised levels of CKMB (McCleod, Neil-Dwyer, Meyer *et al.* 1982). Continuous electro-cardiographic recordings disclosed pronounced sinus tachycardia without any beat variation present in 5 of the 7 patients, superventricular tachycardia (SVT) with heart rates exceeding 150 beats per minute in 4 patients and ECG evidence of significant myocardial ischaemia occurring in 3 patients with ST depressions of 3, 5, and 13 mm respectively.

Ventricular arrhythmias were infrequent with the exception of one patient who demonstrated salvos of ventricular tachycardia on day 1 followed by ST segment depresson on day 3. This patient died suddenly on day 5 after making satisfactory neurological progress. Histological study of the heart showed focal myocardial necrosis in the left ventricle.

This pilot study prompted a larger study which was designed to assess firstly, a possible association between increased sympathetic activity producing cardiac morbidity in acute head injury and secondly, the effect of beta 1 selective blockade with atenolol upon cardiac morbidity.

This study involved 114 haemodynamically stable acute head injury cases who were randomized, double-blind to either placebo or atenolol given intravenously (10 mg six hourly) for 3 days, followed by oral (100 mg once daily) for a further 4 days. Patients were clinically assessed for entry to the trial using the Glasgow coma scale throughout the study, heart rate, and blood pressure was measured every 4 hours, a daily 24-hour double channel ECG recording was performed for up to 5 days and over the 7 days of the study daily measurements of plasma noradrenaline, total creatine kinase (CK) and its myocardial isoenzyme CKMB were done.

As arterial rather than venous concentration of noradrenaline (NA) was felt to be a more sensitive estimate of sympathetic activity in the body as a whole, only the 69 patients who had arterial samples taken will be considered. Plasma noradrenaline levels in many patients were found to be markedly elevated beyond the normal upper limit of 5 n·mol/litre (Fig. 7). There did not appear to be any obvious treatment effect on circulating NA levels. Very high values of mean arterial total CK were found in both groups of patients and no difference was seen between the placebo and treated group. Usually, CKMB is not found in the plasma. A CKMB of greater than 3% of total CK concentration

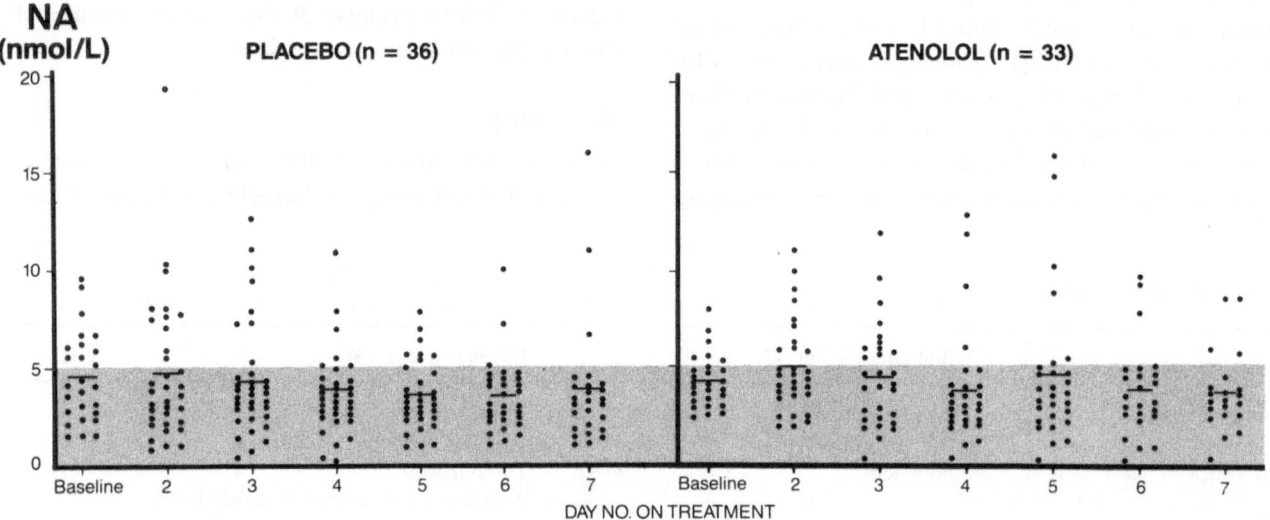

Fig. 7. Arterial noradrenaline (NA) concentrations. Arterial NA levels (daily mean shown as horizontal bar) over 1 week in patients with head injury who received either placebo or atenolol. Shaded area represents normal limits of plasma NA concentration

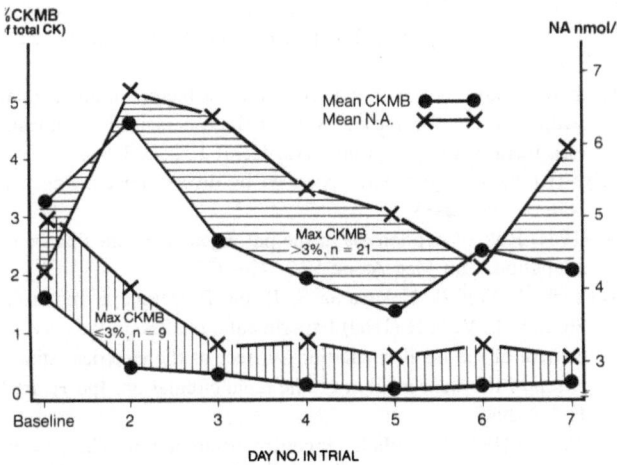

Fig. 8. Mean plasma CKMB and noradrenaline in placebo patients with maximum CKMB >3% and ≤3% of total CK. Relationship over 1 week of plasma NA concentration and CKMB concentration (expressed as percentage of plasma total CK concentration)

is compatible with myocardial damage and over 6% is considered to be evidence of myocardial infarction. While 30% (9 out of 30) of patients in the placebo group had CKMB levels of greater than 3% of total CK, only 7.4% (2 out of 27) in the atenolol group had CKMB of greater that 3% (p < 0.05). CKMB levels > 6% of total CK were detected in 16.7%. ($^5/_{30}$) of pa-

tients receiving placebo and in no patient receiving atenolol (p–0.53).

We examined the relationship between CKMB and plasma noradrenaline in the placebo group of patients to exclude the effect of atenolol. Figure 8 shows the close association between plasma CKMB and noradrenaline concentration in patients on placebo with CKMB levels greater than and less than 3% of total CK. There is a significant correlation (R = 0.46, p = 0.01) between plasma CKMB and NA levels. The incidence of superventricular tachycardia (SVT) is shown in Table 5. The only significant changes was in the reduction of SVT by atenolol (p < 0.0001). The effect of treatment on ST segment and T wave changes is shown in Table 6). Patients who received the atenolol were significantly (p < 0.05) less likely to develop ST segment and T wave changes.

6 hearts were available for detailed examiantion. Focal necrotic lesions were not detected in any of the 4 hearts from the atenolol-treated patients while, in contrast, nectrotic lesions were identified in both of the hearts of the placebo patients.

Conclusion

In demonstrating the stress response in the two conditions, our results show a significant relationship between high catecholamine levels. ECG abnormalities and necrotic myocardial lesions. There was a significant positive correlation between arterial noradrenaline concentration and cardiac damage. These relationships were abolished by beta blockade. The hypothesis remains that the stress response in these conditions is initiated by hypothalamic dysfunction. Finally these studies illustrate that the heart is an excellent monitor of the stress response, particularly as its peformance can be continuously and non-invasively recorded for 24-hour periods.

Table 5. *Incidence of SVT in placebo and atenolol groups (24-hour Holter monitoring)*

Treatment	SVT	
	Day 0 + 1	Days 2–5
Placebo	19	28
Atenolol	10	6
P-value	p < 0.0001	

SVT = Supraventricular tachycardia.

Table 6. *ST segment and T wave changes in placebo and atenolol groups (24-hour Holter monitoring)*

	(n)		ST segment elevation		ST segment depression		T wave inversion		Total changes	
	Day 0 + 1	Days 2 + 5	Day 0 + 1	Days 2–5	Day 0 + 1	Days 2–5	Day 0 + 1	Days 2–5	Day 0 + 1	Days 2–5
Placebo	53	49	0	6	2	4	9	16*	11	26**
Atenolol	51	46	6	5	3	3	6	7	15	15

* p = 0.057 compared with atenolol.

** p = 0.062 compared with atenolol.

References

Crane MG, Harris JL (1976) Effect of aging on renin activity and aldosterone excretion. J Lab Clin Med 87: 947–959

Hippocrates. An English translation by Jones WHS (1923) Vol 1. William Heineman Ltd, Havard University Press

Hoff EC, Kell Jr JF, Hastings N, Sholes DM, Gray EH (1951) Vasomotor, cellular and functional changes produced in the kidney by brain stimulation. J Neurophysiol 14: 317–332

Hunter J (1794) A treatise on the blood, inflammation and gunshot wounds, p 190. Nicol London

James IM (1972) Electrolyte changes in patients with subarachnoid haemorrhage. Clin Sci 42: 179–187

Kerr JH, Corbett JL, Prys-Roberts C, Crampton-Smith A, Spalding JMK (1968) Involvement of the sympathetic nervous system in tetanus. Lancet ii: 236–241

Magoun HW, Ranson SW, Hetherington A (1937) The liberation of adrenin and sympathin induced by stimulation of the hypothalamus. Am J Physiol 119: 615–622

McLeod AA, Neil-Dwyer G, Meyer CHA, Richardson PL Cruickshank JM, Bartlett J (1982) Cardiac sequelae of acute head injury. Br Heart J 47: 221–226

Oka M (1956) Effect of cerebral vascular accidents on the level of 17-hydroxycorticosteroids in plasma. Acta Med Scand 156: 221–226

Raab W (1966) Emotional and sensory stress factors in myocardial pathology. Neurogenic and hormonal mechanisms in pathogenesis, therapy and prevention. Am Heart J 72: 583–664

Selye H (1936) A syndrome produced by diverse nocuous agents. Nature 138: 32–33

— (1944) Role of hypophysis in the pathogenesis of the disease of adapation. Can Med Assoc J 50: 426–433

Takeuchi J, Yagi S, Nakayama S, Ikeoa T, Uchida E Inque G, Shintani F, Veda H (1960) Experimental studies on the nervous control of the renal circulation—Effect of the electrical stimulation of the diencephalon on the renal circulation. Jpn Heart J 1: 288–299

Zanchetti A (1976) Hypothalamic control of circulation. The nervous system in arterial hypertension. Julian S, Esler MD (eds) Thomas Ch C, Springfield, Ill

Correspondence: Department of Neurosurgery, Brook General Hospital, Shooters Hill Road, Woolwich, London SF18 4LW, U.K.

Acta Neurochirurgica, Suppl. 47, 111–113 (1990)

Hyponatraemia and Volume Status in Aneurysmal Subarachnoid Haemorrhage

E. F. M. Wijdicks, M. Vermeulen, and J. van Gijn

Departments of Neurology, University Hospitals Rotterdam and Utrecht, The Netherlands

The aim of this study was to investigate if and why patients with aneurysmal subarachnoid haemorrhage (SAH) and hyponatraemia have a poor outcome. Hyponatraemia is often explained by the syndrome of inappropriate secretion of antidiuretic hormone (SIADH). In that case sustained secretion of antidiuretic hormone is maintained in the face of low osmolality and an expanded extracellular fluid volume, which in turn causes hyponatraemia and natriuresis (Lester and Nelson, 1981). Recently, however, it has been questioned whether inappropriate secretion of ADH really occurs after SAH, since decreased blood volumes have been demonstrated in patients with hyponatraemia, although they fulfilled the laboratory criteria for SIADH (Nelson et al. 1981). This doubt was corroborated by our retrospective review of a consecutive series of 134 patients with SAH.

Hyponatraemia (sodium level of 120 to 134 mmol/l) developed in 44 (33%) of the 134 patients. It was mild (130 to 134 mmol/l) in 18, moderate (125 to 129 mmol/l) in 20, and severe (120 to 124 mmol/l) in 6. In all instances hyponatraemia developed between days 2 and 10 (median 4) after the haemorrhage. Twenty-five patients with hyponatraemia fulfilled the laboratory criteria for SIADH (18% of 134). Twelve patients had other conditions that might have contributed to hyponatraemia, sepsis (2 patients), severe vomiting (3 patients), renal disease (3 patients), Addison's disease (1 patient) and diuretics (3 patients). In 7 cases the cause remained unknown because of incomplete laboratory investigations.

We found the incidence of cerebral infarction after SAH to be significantly higher in patients who developed hyponatraemia (Wijdicks et al. 1985 a).

Table

Serum sodium level	No. of patients		Total
	with infarction	without infarction	
135 mmol/l	19*	71*	90
135 mmol/l	27	17	44
Total	46	88	134

* Significantly different from number in hyponatremic group by Chisquare rest (p < 0.001). From Wijdicks et al. with permission of Annals of Neurology (Wijdicks et al. 1985 a).

We did not find episodes of clinical deterioration that could be attributed to hyponatraemia alone, without clinical or radiological signs of cerebral ischaemia, probably explained by the gradual development of hyponatraemia.

If our patients had a decrease in plasma volume caused by salt wasting, fluid restriction might have aggravated a hypovolemic state, leading to a decrease in cerebral perfusion pressure and cerebral ischaemia.

To elucidate the cause of hyponatraemia, SIADH or cerebral salt wasting, a prospective study was performed on plasma volume, sodium balance and secretion of antidiuretic hormone and renin (Wijdicks et al. 1985 b). In 21 patients with aneurysmal SAH we could demonstrate that the plasma volume decreases in patients who develop hyponatraemia, by comparing two measurements (one on admission, and one after 5 days).

Hyponatraemia was associated with a decrease in plasma volume and a decrease in body weight and preceded by a negative sodium balance in all instances.

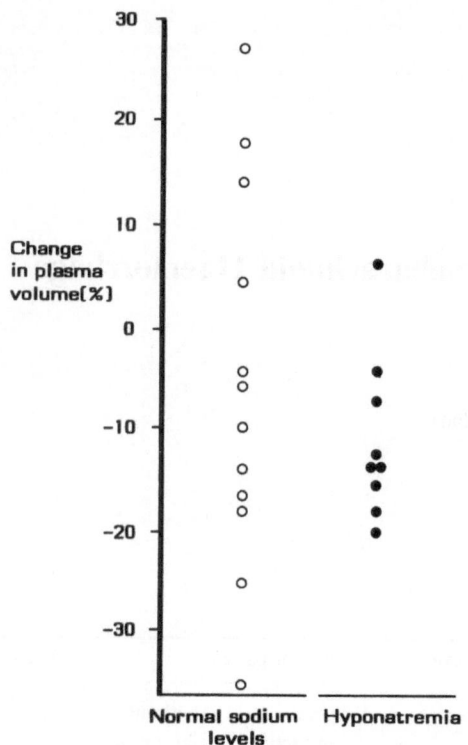

Fig. 1. From Wijdicks *et al.* with permission of the Annals of Neurology

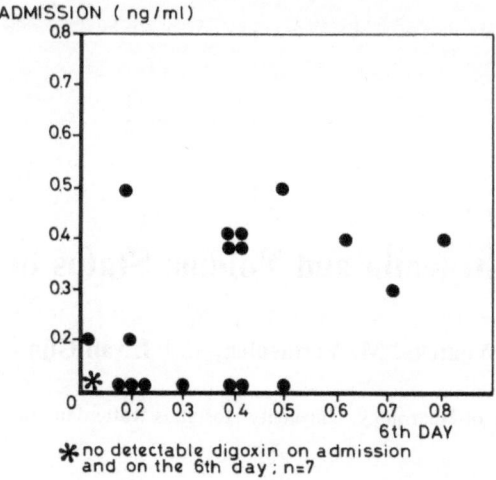

Fig. 2. From Wijdicks *et al.* with permission of the British Medical Journal

Serum vasopressin levels were increased or normal on admission but had decreased at the time of hyponatraemia.

Conversely plasma renin values increased between the two measurements. An additional finding was that plasma volume also decreased considerably in patients with normal serum sodium levels, usually as a result of excessive natriuresis.

All these findings support the concept of cerebral salt wasting and negate the syndrome of inappropriate ADH secretion.

After we had demonstrated that hyponatraemia after SAH is caused by natriuresis, we looked for a substance which possessed natriuretic properties that could be detected in plasma of patients with SAH—the so-called digoxin-like immunoreactive substance. After heating the plasma samples an endogeneous substance cross reacting with antibodies to digoxin was identified in 18 of 25 consecutive patients with SAH (ten of the 25 patients with SAH had detectable levels in both samples—admission and 6th day; 8 in only one sample).

The presence of this substance was associated significantly with an extensive haemorrhage and with a distribution of blood compatible with a ruptured anterior cerebral artery aneurysm and hence suggestive of hypothalamic damage (Wijdicks *et al.* 1987).

In a previous study, a relationship was found in patients with SAH between the development of hyponatraemia and the presence of hydrocephalus on

Ventricular size on the initial CT scan	Hyponatremia	Normal serum sodium	Total
Enlargement of third ventricle			
with enlarged lateral ventricles	12*	14	26
with normal lateral ventricles	8**	7	15
Normal third ventricle			
with enlarged lateral ventricles	2***	1	3
with normal lateral ventricles	22	67	89
Total	44	89	133

*$p_2 = 0.05$ ⎫
**$p_2 = 0.03$ ⎬ Fisher's exact probability test
***N.S.   ⎭

Fig. 3

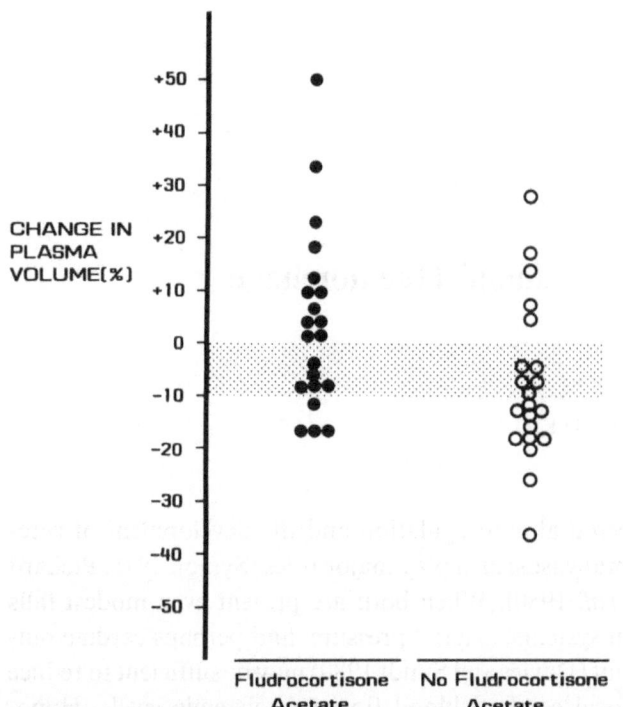

Fig. 4. Plasma volume changes in 21 patients with a ruptured aneurysm treated with fludrocortisone acetate compared with patients without fludrocortisone acetate treatment from our previous study (Mann-Whitney, $p < 0.04$)

admission (Van Gijn *et al.* 1985). A possible explanation was that enlargement of the third ventricle might interfere with hypothalamic function. However, the width of the third ventricle was not measured, as the diagnosis of acute hydrocephalus was based on the width of the lateral ventricles. Therefore, the relationship between hyponatraemia and the size of the third ventricle was investigated separately in a consecutive series of 133 patients, who were seen within 72 hours of aneurysmal haemorrhage.

Hyponatraemia occurred significantly more often in patients with initial enlargement of the third ventricle and lateral ventricles than in patients with a normal ventricular system, and also in patients with enlargement of the third ventricle only (Fig. 3). The relationship between initial enlargement of the third ventricle and hyponatraemia remained after adjustment for the amount of cisternal blood or for mild degrees of hyponatraemia, but not entirely after adjustment for the amount of intraventricular blood (Wijdicks *et al.* 1988 a).

Finally, in order to prevent a negative sodium and

fluid balance we tried the administration of extra sodium solutions or of albumen, but in some of these patients the increased fluid intake was followed by a proportionately increased fluid excretion.

In a pilot study we could demonstrate that fludrocortisone acetate, a mineralocorticoid agent, was a comparatively safe method to increase plasma volume.

In a consecutive series of 39 patients with CT evidence of SAH, fludrocortisone acetate treatment was started on admission. In 21 of these the effect of fludrocortisone acetate on sodium balance and plasma volume could be studied. The plasma volume decreased more than 10% in four of the 21 patients, decreased less than 10% in five and increased in 12 patients. Permanent side-effects did not occur (Wijdicks *et al.* 1988 b).

However, further studies are needed to confirm these results in a comparison with concurrent controls and to demonstrate that these measures result in fewer cerebral infarcts.

References

Lester MC, Nelson PB (1981) Neurological aspects of vasopressin release and the syndrome of inappropriate secretion of antidiuretic hormone. Neurosurg 8: 735–740

Nelson PB, Seif SM, Masoon JC *et al* (1981) Hyponatraemia in intracranial disease. Perhaps not the syndrome of inappropriate secretion of antidiuretic hormone (SIADH). J Neurosurg 55: 938–941

Van Gijn J, Hijdra A, Wijdicks EFM *et al* (1985) Acute hydrocephalus after aneurysmal subarachnoid haemorrhage. J Neurosurg 63: 355–362

Wijdicks EFM, Vermeulen M, Hijdra A, Van Gijn J (1985 a) Hyponatraemia and cerebral infarction in patients with ruptured intracranial aneurysms: is fluid restriction harmful? Ann Neurol 17: 137–140

— — ten Haaf JA *et al* (1985 b) Volume depletion and natriuresis in patients with a ruptured intracranial aneurysm. Ann Neurol 18: 211–216

— — Van Brummelen P *et al* (1987) Digoxin-like immunoreactive substance in patients with aneurysmal subarachnoid haemorrhage. Br Med J 294: 729–732

— van Dongen KJ, van Gijn J, Hijdra A, Vermeulen M (1988 a) Enlargement of the third ventricle and hyponatremia in aneurysmal subarachnoid haemorrhage. J Neurol Neurosurg Psychiatry 51: 516–520

— Vermeulen M, van Brummelen P, van Gijn J (1988 b) The effect of fludrocortisone acetate on plasma volume and natriuresis in aneurysmal subarachnoid hemorrhage. Clin Neurol Neurosurg 90: 209–214.

Correspondence: E. F. M. Wijdicks, M.D., Department of Neurology, University Hospital Utrecht, Box 85500, 3508GA Utrecht, The Netherlands.

Acta Neurochirurgica, Suppl. 47, 114–121 (1990)
© by Springer-Verlag 1990

Blood Volume Measurement Following Subarachnoid Haemorrhage

R. J. Nelson

Wessex Neurological Centre, Southampton General Hospital, Southampton, U.K.

Introduction

The accurate assessment of circulating blood volume by direct clinical observation is notoriously difficult (Irwin *et al.* 1972). Measurements of central venous pressure, pulmonary wedge pressure and cardiac output may provide useful additional information about the cardiovascular status of patients, but involve invasive techniques with attendant risks, and require intensive care facilities (Finn *et al.* 1986, Pritz 1984 a, b). In contrast, blood volume measurements using radioisotopes are bedside procedures. However, for several reasons they have not been widely adopted in routine surgical practice. They require certain laboratory facilities and technical support and tend to be less accessible to the clinician who, as a result, remains unfamilar with their use. They are relatively time consuming to perform and unless sufficient care is taken significant methodological errors may occur. Finally, and perhaps most importantly, the problems of standardizing or predicting normal values may seriously hamper the interpretation of results.

Despite these practical difficulties recent blood volume studies have added to our knowledge of the metabolic responses to intracranial pathology. They have highlighted the relationship which exists between delayed cerebral ischaemia and systemic hypovolaemia following spontaneous subarachnoid haemorrhage (SAH) and have contributed to our understanding and management of hyponatraemia. This chapter reviews the published literature and suggests how blood volume measurement can contribute to future clinical research.

Hypovolaemia

The pathogenesis of delayed cerebral ischaemia following a SAH remains uncertain but the impairment of cerebral autoregulation and the development of cerebral vasospasm play major roles (Symon 1978, Pickard *et al.* 1980). When both are present even modest falls in systemic arterial pressure, and perhaps cardiac output (Davies and Sundt 1980) may be sufficient to reduce focal cerebral blood flow to ischaemic levels. Hypovolaemia reduces cardiac output and exposes patients to increased risks of hypotension and cardiovascular instability.

Maroon and Nelson (1979) were the first to document a significant decrease in the red cell volume (RCV) and total circulating blood volume (TCBV) of 15 nonselected SAH patients compared with six unmatched neurosurgical patients without intracranial disease. The measurements were made an average of ten days after the SAH using [51]chromium-labelled autologous red blood cells and radio-iodinated human serum albumin (RISA). They failed to detect a significant change in plasma volume (PV). Limited information was provided regarding the relationship between hypovolaemia and either the neurological status of their patients, the presence of cerebral vasospasm or the development of ischaemic neurological deficits.

Since then a number of prospective studies have demonstrated a correlation between hypovolaemia and the development of cerebral ischaemia. Kudo *et al.* (1981) reported three cases from a series of 45 patients in whom PV was monitored both pre- and post-operatively using the RISA technique. In each case ischaemic symptoms occurred in the presence of hypovolaemia and were reversed by blood volume expansion. They calculated the RCV of their patients from their PV and venous haemocrits (vide infra). These cases were studied at least 14 days after their haemorrhage and it was a reduction in RCV rather than PV that the investigators considered to be important.

Solomon *et al.* (1984) studied 25 unselected SAH patients using [51]chromium-labelled autologous red blood cells. Whilst there was no significant reduction in the blood volumes of the seven males as a group, the RCV and TCBV of the 18 females were significantly lower than corresponding control values. Fifteen of the patients had angiographic evidence of vasospasm and seven of these (six of whom were females) developed neurological deficits attributed to cerebral ischaemia. The RCV and TCBV of the six females with "symptomatic" vasospasm were significantly lower than those with "asymptomatic" spasm.

Several important questions remain unanswered. Which patients are at greatest risk of developing hypovolaemia? What is the time-course of hypovolaemia and can it be prevented? Is hypovolaemia following a SAH a non-specific phenomenon or is it the result of a single pathophysiological disturbance? Maroon and Nelson attributed hypovolaemia to the effects of bedrest, the supine diuresis, pooling in the peripheral vascular beds, negative nitrogen balance, decreased erythropoiesis, and iatrogenic blood loss. These largely catabolic processes would be expected to result in a gradual contraction of blood volume, particularly of RCV. In the studies described above a consistent finding in hypovolaemic patients was a low RCV occurring at least a week after the haemorrhage. Similar circulatory changes have been observed in other clinical and experimental studies. Trauma patients who have been resuscitated following blood loss with crystalloid and colloid solutions suffer a 30–40% reduction in PV and RCV, exposing them to hypotension and even cardiac arrest during anaesthesia (Biron *et al.* 1972). In dogs blood loss replaced by isotonic sodium chloride results in a 10% fall in TCBV, a 50% fall in RCV, but no change in PV (Valeri *et al.* 1986). Similar, if less profound changes in blood volume would be anticipated in poor-grade SAH patients who typically received fluid replacement with isotonic or hypotonic saline solutions whilst awaiting surgery. Sundt's assertion that those patients developing cerebral ischaemia should be treated with "less drug and more blood" appears justified (Sundt and Whisnant 1978). Hypovolaemia due to a reduction in RCV should be considered a potential problem in any patient undergoing delayed aneurysm surgery.

Recent evidence suggests that hypovolaemia can also occur in the first week after a SAH due to a reduction in PV. Wijdicks *et al.* (1985 a) measured PV in 21 patients within 48 hours of admission and again six days after their haemorrhage. PV decreased by more than 10% in 11 patients. Fluid and sodium balance studies performed at the same time implicated a negative sodium balance due to an excessive natriuresis in the majority of the hypovolaemic patients.

It seems likely that the early development of hypovolaemia will be reflected by changes in PV rather than RCV. Under these circumstances the prevention of hypovolaemia will depend on measures directed towards maintaining PV.

Hyponatraemia

Some 25% of patients with serious intracranial pathology will develop hyponatraemia (Fox *et al.* 1971). Hyponatraemia is frequently associated with, and may be responsible for neurological deterioration. The pathophysiology of hyponatraemia in these patients is still not fully understood. In 1950 Peters *et al.* suggested that the problem was due to excessive renal excretion of sodium and coined the term "cerebral salt wasting". The primary disturbance was considered to be a centrally mediated influence on the ability of the kidney to reabsorb sodium from the proximal tubule resulting in a secondary reduction in extracellular fluid volume (Cort 1954). There was no evidence of glucocorticoid or mineralocorticoid deficiency and treatment with salt replacement was advised.

Attitudes changed in the early 1960's after the report of Schwartz *et al.* (1957) describing two patients with hyponatraemia and bronchogenic carcinomas which they attributed to a syndrome of inappropriate secretion of antidiuretic hormone (SIADH). The essential features of the syndrome—progressive hyponatraemia, urinary sodium loss, urine persistently hypertonic to plasma, resistance to correction by hypertonic saline but response to fluid restriction—had already been observed in normal subjects given pitressin and water. It is important to note that a reduction in blood volume was not a prerequisite of SIADH as originally described. In many subsequent studies blood volume was found to be normal (Kaye 1966, Ivy 1961). Although an increase in total body water can occur in patients with SIADH, this rarely amounts to more than 3–4 litres and is not accompanied by oedema (Bartter and Schwartz 1967). Inappropriate ADH secretion in neurological patients was then widely reported such that SIADH became accepted as the principle cause of hyponatraemia following SAH and head injury and fluid restriction was adopted as the treatment of choice (Fox *et al.* 1971, Doczi *et al.* 1981, 1982).

An antidiuretic phase is a predictable component

of the normal response to stress and it has been suggested that it is an adaptive mechanism to prevent dehydration in the ill or injured (Zimmermann 1965, Moore and Ball 1952). The response is mediated by high levels of vasopressin (ADH) and has been demonstrated following craniotomy and head injury (McLaurin *et al.* 1960 a, b, 1975). Vasopressin release may be prolonged by the specific features of intracranial disease including hypothalamic-pituitary disturbances and raised intracranial pressure (Jenkins *et al.* 1980). Bouzarth and Shenkin (1982) warned against the uncritical use of term "inappropriate" ADH secretion and claimed that "cerebral hyponatraemia" was an iatrogenic phenomenon largely due to excessive administration of fluids to patients in a water-retaining state.

A further examination of SIADH in neurological patients by Nelson *et al.* (1981) using blood volume measurements turned the preceeding arguments full circle. They reported on a series of 12 unselected neurosurgical patients (8 SAH, 2 head injury, 2 craniotomy) who fulfilled the laboratory criteria for SIADH, becoming hyponatraemic, on average, on the 10th day of their illness (range 4–16 days). Blood volume was determined using ^{51}Cr-labelled RBCs and RISA within 48 hours of making the diagnosis of SIADH and the results, expressed in terms of body weight, were compared with the blood volumes of six neurosurgical patients without intracranial disease. The mean blood volumes of the hyponatraemic patients were found to be significantly lower than those of the controls, although blood volume was expanded in two patients. The authors felt that the reduced blood volumes were more in keeping with the original concept of cerebral salt wasting, whilst admitting that there might be a spectrum of abnormalities in the hyponatraemic, natriuretic patient ranging from salt wasting to SIADH.

They cautioned against the use of fluid restriction alone in treating hyponatraemia, recognizing the risks to the cerebral circulation of adding further dehydration to a hypovolaemic, hyponatraemic state. These risks were emphasized in a retrospective clinical study of hyponatraemia and cerebral infarction in 134 SAH patients published in 1985 (Wijdicks *et al.* 1985 b). Of 44 patients who became hyponatraemic 25 fulfilled the criteria for SIADH. Cerebral infarction developed in 27 of the 44 hyponatraemic patients and in 19 of the 90 normonatraemic patients (p < 0.001) and cerebral infarctions were more commonly fatal in the hyponatraemic patients (p < 0.01). Cerebral infarction occurred in 21 of the 26 patients treated by fluid restric-

tion compared with 6 of 18 patients with a normal fluid intake (p = 0.004). It appeared that hyponatraemia preceeded cerebral infarction and was not simply a epiphenomenon.

The study of relationships between a natriuresis, hyponatraemia, and hypovolaemia in SAH patients is made difficult by the complexities of controlling salt and water balance in critically ill patients and the lack of pre-morbid data to allow interpretation of results. To overcome these problems Nelson's group turned to a monkey model of SAH to examine the incidence of hyponatraemia and natriuresis together with changes in ADH secretion and salt and water balance (Nelson *et al.* 1984). PV was determined using the RISA technique before and after the SAH and at the nadir of serum sodium content (mean 4.4 days). Seven of 9 SAH animals developed a negative sodium balance and natriuresis and became hyponatraemic, the falls in PV in this group were less than in sham-operated animals and neither were statistically different from pre-operative values. Plasma vasopressin levels rose transiently on the first post-operative day. The authors concluded that changes in salt balance were mainly responsible for the development of hyponatraemia and that the release of a brain natriuretic factor or alteration in the neural control of the kidney should be considered possible causes of a primary natriuresis.

The work of Wijdicks *et al.* (1985) showed that a negative sodium balance (12 out of 21 SAH patients) is likely to be associated with both hyponatraemia (8 of 9 patients) and with hypovolaemia (10 of 12 patients). Nevertheless, there were considerable individual variations between patients indicating that the pathophysiological disturbances were not of a uniform pattern. Thus four patients without hyponatraemia were in negative sodium balance and three hyponatraemic patients were not hypovolaemic.

In summary:

1) Hypovolaemia following a SAH may occur as a result of reductions in both PV and RCV. A consideration of the time-course of the processes involved suggests that early hypovolaemia will be due to changes in PV and will be followed by a gradual catabolic reduction in RCV. These observations dictate to a certain extent the timing and type of blood volume measurement to be used in future studies.

2) Although a natriuresis and negative sodium balance proceed the development of hypovolaemia and hyponatraemia in many patients, it is most unlikely

that there is a "unified" metabolic response to a SAH and some attention must be paid to identifying the different types of response.

3) To do this changes in blood volume must be related, to each individual patient's normal, physiological, circulating blood volumes.

Methodological Aspects of Blood Volume Measurement

Modern isotope dilution methods have superceded all others for the direct measurement of blood volume. The techniques employed do not vary greatly between laboratories. Internationally agreed guidelines have been set out (ICSH 1973) and the potential sources of laboratory error are well described elsewhere (Wright et al. 1975). However, several practical matters deserve mention.

The gold standard of blood volume measurement is the simultaneous determination of both PV and RCV. This is usually achieved with radioiodinated (^{125}I or ^{131}I) human serum albumin (RISA) and autologous red blood cells labelled with sodium radiochromate (^{51}Cr) or sodium pertechnetate (^{99}Tc), which, because of its short half-life, is better suited to serial measurements of RCV. The additional resources required to perform simultaneous measurements make single volume measurements attractive. It is possible to estimate TCBV from either PV or RCV using the peripheral venous haematocrit (Hv), having first corrected for the difference between the venous and total body haematocrit. A difference exists because the haematocrit of blood varies throughout the circulation, capillary blood having the lowest haematocrit and venous blood the highest. Total body haematocrit (a theoretical figure which can only be derived from simultaneous measurements of RCV and PV) represents the "true" mean haematocrit and is related to venous haematocrit by the body/venous haematocrit ratio (Hb/Hv, sometimes referred to as the f-cell ratio). The ratio is usually taken to be 0.9 and thus the formulae for calculating TCBV become:

$$\text{TCBV} = \frac{RCV}{0.9\,\text{Hv}} \text{ or } \frac{PV}{1-0.9\,\text{Hv}}$$

It has been generally assumed that the Hb/Hv ratio does not change greatly between individuals or groups of patients, but this is not the case. Wide variations occur in patients with haematological or circulatory disturbances e.g. spleenomegaly, severe anaemia or shock (Loria et al. 1962), and might be expected in any patient with hypovolaemia. We found a significant re-

duction in the Hb/Hv ratio of 10 unselected SAH patients compared with 10 matched controls from 0.91 to 0.87 and speculated that this might be due to a reduction in central venous blood volume (Nelson et al. 1987). Kirsch et al. (1971) found similar changes in 29 hospitalized patients and concluded that TCBV should be calculated using a Hb/Hv ratio of 0.86 rather than 0.91. In our series the mean errors of calculating TCBV from either RCV or PV and the appropriate Hb/Hv ratio, 0.87, were 2.84% and 1.76% respectively.

The accuracy of isotope dilution techniques depends upon complete and homogeneous mixing of the isotope marker in the space to be measured. Ideally the marker should remain entirely within the space or should leave it at a predictable and constant rate. Albumin is continuously lost from the circulation to the lymphatic and reticuloendothelial systems (Rossing 1967, Rothschild 1955) with the result that the uncorrected albumin space measured with RISA is always greater than the true PV. This problem may be overcome by counting the activity of a number of dilutional samples (say 10, 20 and 30 minutes) and extrapolating back to time-zero, giving a measure of perfect or instantaneous mixing of RISA in the plasma. Alternatively a single sample may taken at 10 minutes and a correction made for the loss of RISA. (The ICSH recommends a factor of 1.015 which assumes that 1.5% of the RISA is lost from the circulation in 10 minutes).

Opinions vary as to whether single sample determinations of PV need to be corrected and even if they are desirable in the first place. In the absence of hypovolaemic shock or congestive cardiac failure the mixing of RISA in the albumin space is complete by 10 minutes and the loss of activity is probably less than 1% (Noble and Gregersen 1946). Using 15 and 30 minute samples we showed no difference between the rate of loss of activity in SAH patients and controls. From a practical point of view, small differences, of the order of 2–3%, occur when PV is calculated either from a 10-minute equilibration sample or by back-extrapolation (Tarazi et al. 1968). The later technique involves multiple venepunctures to obtain stasis-free blood and additional iatrogenic blood loss which may be undesirable when serial PV determinations are contemplated.

The problem of standardizing and predicting normal blood volumes and thus interpreting individual results far outweighs other methodological problems. The commonest method of presenting blood volume data is in terms of body weight. Normal values for RCV for adult males and females are 30 ml/kg (25–35

95% confidence limits) and 25 ml/kg (20–30) respectively. The values for PV are roughly the same in men and women, 40 ml/kg, but the 95% confidence limits are even greater. These large variations in normal values reflect differences in body composition. Since fat is relatively avascular blood volume is much more closely correlated with lean body mass than body weight. The obese elderly patient will have much lower blood volumes expressed in terms of body weight than say a fit athletic young man. Even in healthy individuals blood volume varies with physical training, activity, nutritional state, body position, menstrual cycle, season and altitude. Given these observations the most stable determinant of normal blood volume would appear to be lean body mass. Because there is no direct or simple way of measuring lean body mass attempts have been made to predict normal blood volumes using formulae based on combinations of weight, height, and surface area but the results are only slightly better than predictions based on weight alone (Retzlaff *et al.* 1969, Hurley 1975, Wennesland *et al.* 1959, Naddler *et al.* 1962).

In the SAH research described above the difficulties of blood volume prediction were partly overcome by either comparing the blood volumes of patient groups with controls, or by following serial changes in the blood volumes of each patient. These approaches only allow a qualitative assessment of blood volume status which may not be sufficient to define different pathophysiological subgroups. We have attempted to improve the accuracy of blood volume prediction in SAH patients by re-examining the use of total body water (TBW) measurements, described in detail by Moore *et al.* (1963). TBW is the major constituent of lean body mass and as such is closely related to normal blood

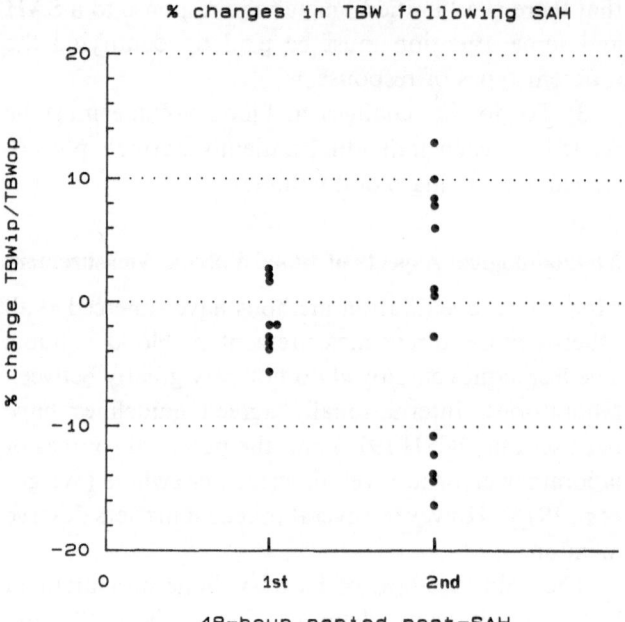

Fig. 1

volume. The prediction of normal blood volumes using TBW as an index of lean body mass circumvents many of the drawbacks of other methods and has the advantage of being a simple and robust measurement.

We studied 32 consecutive patients admitted within 96 hours of a SAH, confirmed by computerized tomographic brain scanning and where necessary lumbar puncture. The TBW, RCV and PV were measured using tritiated water, ^{125}I-RISA and ^{51}Cr-labelled RBCs. Cerebral aneurysms were identified angiographically in 29 patients. In the remaining 3 patients the distribution of blood in the sylvian or anterior interhemispheric fissures made the presence of an aneurysm likely but their clinical condition precluded further investigation.

Table 1. *Summary of regression equations*

Independent variable	Dependent variable	Regression equation	Correlation coefficent
TBWop	TCBVop =	− 0.08 + 0.10 TBWop	r = 0.96
TBWop	PVop =	0.04 + 0.064 TBWop	r = 0.93
TBWop	RCVop =	− 0.11 + 0.04 TBWop	r = 0.92
BWt	TCBVop =	0.16 + 0.056 BWt	r = 0.58
BWt	PVop =	0.08 + 0.036 BWt	r = 0.59
BWt	RCVop =	0.09 + 0.020 BWt	r = 0.52

TBWop = total body water measured in outpatients at 6 months.
TCBV = total circulating blood volume.
PV = plasma volume.
RCV = red cell volume.
BWt = outpatient body weight.

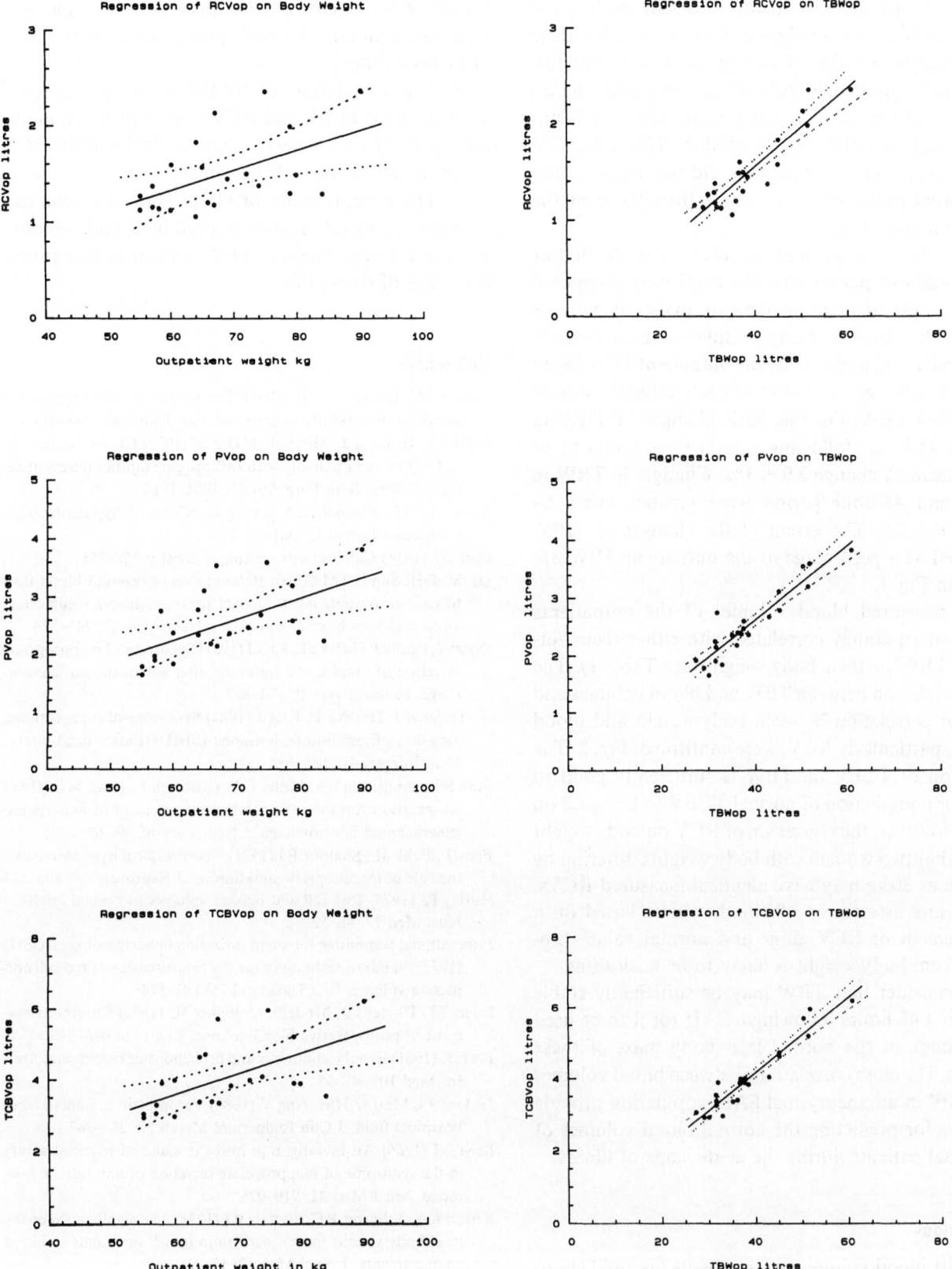

Fig. 2

Twenty-two of the 32 patients who had made good recoveries were reinvestigated 6 months after their haemorrhages. Of the 10 patients who were not followed up 6 had died, 3 declined the outpatient studies for personal reasons and one patient who was hemiparetic and immobile was excluded. The admission characteristics of these patients did not suggest that their initial prognoses were worse than those of the patients followed up.

The TBW measurements made either in the first or second 48-hour period after the SAH were compared with the outpatient measurements made six months later. In the absence of any complications or changes in general health and activity the outpatient TBWs were assumed to be representative of each patient's normal pre-morbid TBW. On this basis changes in TBW in the first 48 hours following a SAH were found to be small, mean % change $2.9 \pm 1\%$. Changes in TBW in the second 48-hour period were greater, mean % change 9 ± 2.5. The extent of the changes in TBW, expressed as a percentage of the outpatient TBW are shown in Fig. 1.

The measured blood volumes of the outpatients were then separately correlated with either their outpatient TBW or their body weight (see Table 1). The close correlation between TBW and blood volumes and the poor correlation between body weight and blood volume, particularly RCV, were confirmed, Fig. 2. The regression of TCBV on TBW is sufficiently good to allow for a prediction of normal TCBV to be based on TBW. However, the regression of RCV on body weight is such that two women with body weights differing by as much as 30 kg may have identical measured RCVs. Clearly any assessment of blood volume based on a measurement of RCV alone and normal values predicted from body weight is likely to be misleading.

We consider that TBW may be sufficiently stable in the first 48 hours following a SAH for it to be used as an index of the normal lean body mass of these patients. The close correlations between blood volumes and TBW in an aneurysmal SAH population provide the basis for predicting the normal blood volumes of individual patients during the acute stage of illness.

Conclusions

1. Whilst blood volume measurements are unlikely to become routine investigations in SAH patients they should be considered as part of any detailed metabolic study or investigation of systemic circulatory changes.

2. Centres adopting an aggressive approach to early surgery in poor grade patients may wish to supplement their usual monitoring with pre-operative assessment of blood volume.

3. The calculation of TCBV from simultaneous measurements of PV and RCV is least prone to methodological error. The assessment of TCBV from RCV alone should be viewed cautiously.

4. The interpretation of blood volume results can be improved by taking into account body composition. The early measurement of TBW appears to be a promising way of doing this.

References

Bartter FC, Schwartz WB (1967) The syndrome of inappropriate secretion of antidiuretic hormone. Am J Med 42: 790–806

Biron PE, Howard J, Altschule MD *et al* (1972) Chronic deficits in red-cell mass in patients with orthopaedic injuries (stress anaemia). J Bone Joint Surg Am 54: 1001–1014

Bouzarth WF, Shenkin HA (1982) Is "Cerebral hyponatraemia" iatrogenic? Lancet i: 1061–1062

Cort JH (1954) Cerebral salt wasting. Lancet i: 752–754

Davies DH, Sundt TM (1980) Relationship of cerebral blood flow to cardiac output, mean arterial pressure, blood volume and alpha and beta blockade in cats. J Neurosurg 52: 745–754

Doczi T, Bende J, Huszka E, Kiss J (1981) Syndrome of inappropriate secretion of antidiuretic hormone after subarachnoid haemorrhage. Neurosurgery 9: 394–397

— Tarjanyi J, Huszka E, Kiss J (1982) Syndrome of inappropriate secretion of antidiuretic hormone (SIADH) after head injury. Neurosurgery 10: 685–688

Finn SS, Stephensen SA, Miller CA, Drobnich L, Hunt WE (1986) Observations on the perioperative management of aneurysmal subarachnoid haemorrhage. J Neurosurg 65: 48–62

Fox JL, Falik JL, Shalour RJ (1971) Neurosurgical hyponatraemia: the role of inappropriate antidiuresis. J Neurosurg 34: 506–514

Hurley PJ (1975) Red cell and plasma volumes in normal adults. J Nucl Med 17: 46–52

International committee for standardization in haematology (ICSH) (1973) Standard techniques for the measurement of red cell and plasma volume. Br J Haematol 25: 801–814

Irwin TT, Hayter CJ, Modgill, Golligher JC (1972) Clinical assessment of postoperative blood volume. Lancet ii: 446–448

Ivy HK (1961) Renal sodium loss and bronchogenic carcinoma. Arch Int Med 108: 47–55

Jenkins JS, Mather HM, Ang V (1980) Vasopressin in human cerebrospinal fluid. J Clin Endocrinol Metab 50: 364–367

Kaye M (1966) An investigation into the cause of hyponatraemia in the syndrome of inappropriate secretion of antidiuretic hormone. Am J Med 41: 910–926

Kirsch KA, Johnson RF, Gorten RJ (1971) The significance of the total body venous haematocrit ratio in measurements of blood compartments. J Nucl Med 12: 17–21

Kudo T, Suzuki S, Iwabuchi T (1981) Importance of monitoring the circulating blood volume in patients with cerebral vasospasm after subarachnoid haemorrhage. Neurosurgery 9: 514–520

Loria A, Sanchez-Medal L, Kauffer N, Quintanar E (1962) Relationship between body haematocrit and venous haematocrit in normal splenomegalic and anemic states. J Lab Clin Med 60: 396–408

Maroon JD, Nelson PB (1979) Hypovolaemia in patients with subarachnoid haemorrhage: Therapeutic implications. Neurosurgery 4: 223–226

McLaurin RL, King L, Tutor FT, Knowles Jr H (1960 a) Metabolic response to intracranial surgery. Surg Forum 10: 770–773

— — Elam EB, Buddle RB (1960 b) Metabolic response to craniocerebral trauma. Surg Gynec Obstet 110: 282–288

— — (1975) Metabolic effects of head injury. In: Vinken PJ, Bruyn GW (eds) Handbook of clinical neurology, vol 23: Injuries of the brain and skull, pp 109–131

Moore FD, Ball MR (1952) The metabolic response to surgery. Ch C Thomas, Springfield Ill

— Olesen JD, McMurrey H, Parker V, Ball MR, Boyden C (1963) The body cell mass and its supporting environment: body composition in health and disease. WB Saunders, Philadelphia

Naddler SB, Hidalgo JU, Bloch T (1962) Prediction of blood volume in normal human subjects. Surgery 51: 224–232

Nelson RJ, Roberts J, Ackery DM, Pickard JD (1987) Measurement of total circulating blood volume following subarachnoid haemorrhage: methodological aspects. J Neurol Neurosurg Psychiatry 50: 1130–1135

Nelson PB, Seif SM, Maroon JC, Robinson AG (1981) Hyponatraemia in intracranial disease: perhaps not the syndrome of inappropriate secretion of antidiuretic hormone (SIADH). J Neurosurg 55: 938–941

— — Gutai J, Robinson AG (1984) Hyponatraemia and natriuresis following subarachnoid haemorrhage in a monkey model. J Neurosurg 60: 233–237

Noble RP, Gregersen MI (1946) Mixing time and disappearance of T-1824 in shock. J Clin Invest 25: 158–171

Peters JP, Welt LG, Sims EAH et al (1950) A salt-wasting syndrome associated with cerebral disease. Trans Assoc Am Physicians 63: 57–64

Pickard JD, Matheson M, Patterson J, Wyper D (1980) Prediction of late ischaemic complications after cerebral aneurysm surgery by the intraoperative measurement of cerebral blood flow. J Neurosurg 53: 305–308

Pritz MB (1984 a) Monitoring cardiac function and intravascular volume in neurosurgical patients. Neurosurgery 15: 775–780

— (1984 b) Treatment of cerebral vasospasm. Usefulness of Swan-Ganz Catheter monitoring of volume expansion. Surg Neurol 21: 239–244

Retzlaff JA, Tause WN, Kieley JM et al (1969) Erythrocyte volume, plasma volume, and lean body mass in adult men and women. Blood 33: 649–661

Rossing N (1967) The normal metabolism of ^{131}I-labelled albumin in man. Clin Sci 33: 593–602

Rothschild MA, Bowman A, Yalow RS, Berson SA (1955) Tissue distribution of ^{131}-labelled human serum albumin following intravenous administration. J Clin Invest 34: 1354–1358

Schwartz WB, Bennet W, Curelop S, Bartter FC (1957) A syndrome of renal sodium loss and hyponatraemia probably resulting from inappropriate secretion of antidiuretic hormone. Am J Med 23: 529–542

Solomon RA, Post KD, McMurty III JG (1984) Depression of circulating blood volume in patients after subarachnoid haemorrhage: Implications for the management of symptomatic vasospasm. Neurosurgery 15: 354–362

Sundt TM, Whisnant JP (1978) Subarachnoid haemorrhage from intracranial aneurysms: Surgical management and natural history. N Engl J Med 229: 116–122

Symon L (1978) Disordered cerebrovascular physiology in aneurysmal subarachnoid haemorrhage. Acta Neurochir (Wien) 41: 7–22

Tarazi RC, Frohlich ED, Dustan HP (1968) Plasma volume in men with essential hypertension. N Engl J Med 278: 762–765

Valeri CR, Donahue K, Feingold HM, Cassidy GP, Altschule MD (1986) Increase in plasma volume after the transfusion of washed erythrocytes. Surg Gynec Obstet 162: 30–36

Wennesland R, Brown E, Hopper J et al (1959) Red cell, plasma and blood volume in healthy mean measured by radiochromium (^{51}Cr) cell tagging and haematocrit: influence of age, somatotype and habits of physical activity on the variance after regression of volumes to height and weight combined. J Clin Invest 38: 1065–1077

Wijdicks EFM, Vermeulen M, ten Haaf JA, Hijdra A, Bakker WH, van Gijn J (1985 a) Volume depletion and natriuresis in patients with a ruptured intracranial aneurysm. Ann Neurol 18: 211–216

— — Hijdra A, van Gijn J (1985 b) Hyponatraemia and cerebral infarction in patients with ruptured cerebral aneurysms: Is fluid restriction harmful? Ann Neurol 17: 137–140

Wright RR, Tono M, Pollcove M (1975) Blood volume. Sem Nucl Med 5: 63–78

Zimmermann B (1965) Pituitary and adrenal function in relation to surgery. Surg Clin N Amer 45: 299–315

Correspondence: R. J. Nelson, F.R.C.S., Senior Registrar in Neurosurgery, Wessex Neurological Centre, Southampton General Hospital, Tremona Road, Southampton, U.K.

Acta Neurochirurgica, Suppl. 47, 122–126 (1990)

Central Neuroendocrine Control of the Brain Water, Electrolyte, and Volume Homeostasis

T. Dóczi, F. Joó, and **M. Bodosi**

Department of Neurosurgery, University Medical School Szeged, Hungary, and Laboratory for Molecular Neurobiology, Biological Research Center, Szeged, Hungary

Background and Review of the Literature

In 1981 Marcus Raichle put forward the hypothesis that a central neuroendocrine system regulates the brain ion and water homeostasis[38]. In this presentation I would like to summarize briefly the available data—including our own results—in support of this hypothesis. The hypothesis supposes that three cell groups (brain capillary endothelial cells, secretory cells of the choroid plexus, and astroglia) regulate the internal ionic environment of the brain[38]. A unique element of this hypothesis is that the regulation of the ion and volume homeostasis of the brain is orchestrated by a central neuroendocrine system capable of affecting all three cell types[38].

What is meant by central as opposed to peripheral with respect to the neuroendocrine system is well demonstrated by considering vasopressin (AVP). Schultz *et al.* (1977), Brownfields and Kozlowski (1977) and Buijs *et al.* (1978) showed that the plasma and cerebrospinal AVP originate from different sources[4, 5, 43]. Further evidence for the existence of a separate central vasopressin system is provided by the inability of AVP to cross the blood-brain barrier and blood-cerebrospinal fluid (CSF) barriers[48]. On the other hand, the secretion of AVP into the CSF may be influenced by the same stimuli that govern its secretion into the blood[51]. According to Rodriguez (1976), the CSF and interstitial fluid (ISF) flows should serve as pathways in neuroendocrine integration[6, 7, 40]. His generally accepted concept is strongly supported by the recent findings of Rennels *et al.* (1985) of dynamic interactions between the CSF and the ISF of the nervous tissue[39].

With a protein tracer administered into the ventricle, they were able to prove that a "paravascular" fluid circulation exists in the mammalian brain[39]. They demonstrated that solutes in the CSF have access to the extracellular space throughout the brain within a matter of minutes (2–4 minutes), via fluid pathways along the intraparenchymal vasculature[39].

The activity of the secretory cells of the choroid plexus may be under neuroendocrine control[3, 24–26, 32, 33, 42, 52]. Furthermore, nerve fibers are present on the secretory cells and on the capillaries of the choroid plexus[52]. These fibers may originate from the peripheral sympathetic chain and from the paraventricular and supraoptic nuclei[4, 24–26, 32]. Specific neurotransmitter and hormone receptors are present on the secretory cells of the choroid plexus[33, 42]. Sympathetic nerve stimulation and exposure of the choroid plexus to AVP and acetylcholine not only alter CSF formation rate but also produce ultrastructural changes in the affected cells[13, 24, 43]. Liszczak *et al.* (1987) showed that fluid transport was enhanced in the epithelium of the choroid plexus in response to AVP in the CSF[27]. Steardo and Nathanson (1987) found receptors and second messenger systems in the secretory cells of the choroid plexus for atriopeptin, a cardiac peptide involved in peripheral water regulation[14, 21, 30, 31, 35, 44, 49].

The role of astroglia in brain extracellular fluid and ionic homeostasis has been studied extensively. Certain observations suggest that the glial cell activity may be under neuroendocrine control. Kimelberg *et al.* (1978) showed that the activities of glial Na, K-ATP-ase and carbonic anhydrase increase in the presence of norepinephrine[22]. Harik *et al.* (1979) and MacKenzie *et al.*

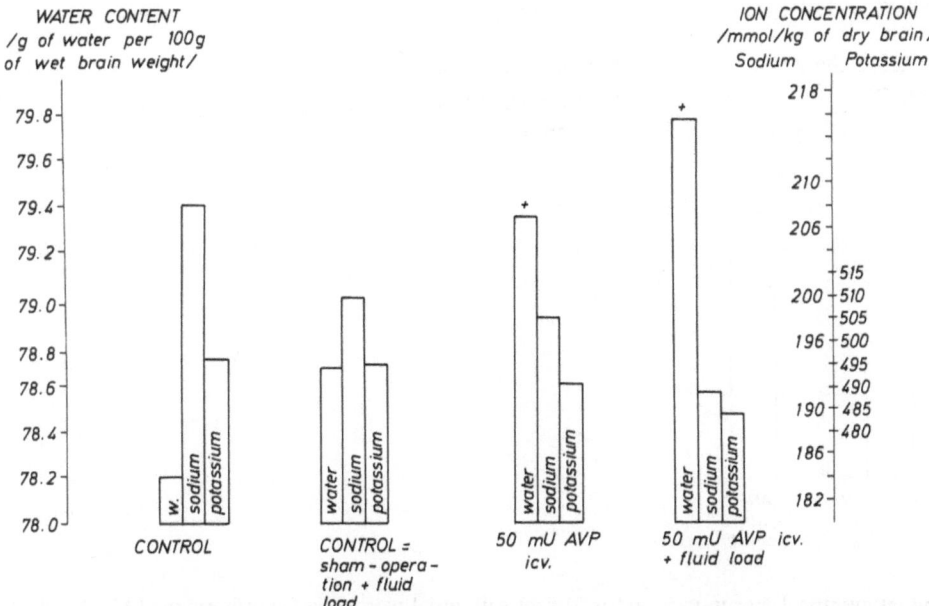

Fig. 1. Brain water content and ion concentrations after central (intraventricular) administration of vasopressin (AVP). Fluid load = intravenous infusion of 2.5% dextrose in water to a volume of 7.5% of the body weight. icv. = intraventricular. + = statistically different from control; p < 0.05

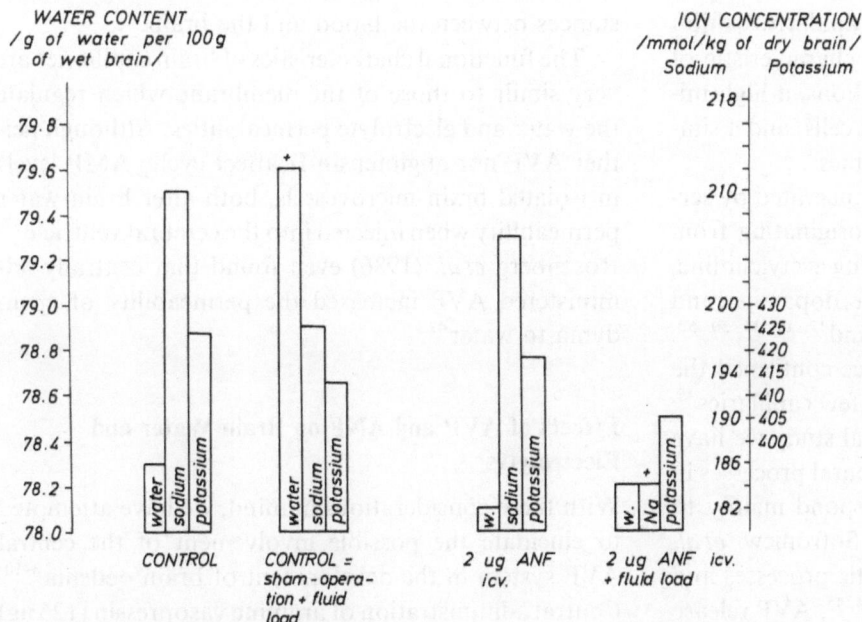

Fig. 2. Changes in brain water content and electrolytes after central administration of atrial natriuretic factor/hormone (ANF). + = statistically different from control; p < 0.05. Fluid load = intravenous 2.5% dextrose in water to a volume of 20% of the body weight. icv. = intraventricular

(1979) also revealed the modulatory role of norepinephrine upon glial cell enzyme activiy [15, 16, 29].

The brain capillary endothelium is considered to be a major component in the regulation of the internal environment of the brain, because of its very large surface area relative to that of the choroid plexus (5000 : 1) and its very close contact with the neuropil [3]. Cserr et al. (1981) suggest that the brain capillaries have the potential to play a very dynamic role in the regulation of brain water and electrolyte permeability, in

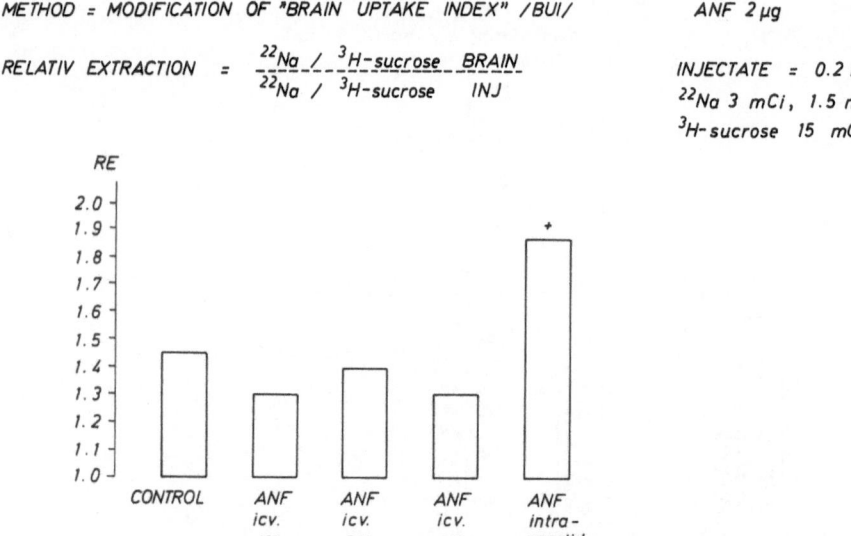

METHOD = MODIFICATION OF "BRAIN UPTAKE INDEX" /BUI/   ANF 2 µg

$$RELATIV\ EXTRACTION\ =\ \frac{^{22}Na\ /\ ^{3}H\text{-}sucrose\ \ BRAIN}{^{22}Na\ /\ ^{3}H\text{-}sucrose\ \ INJ}$$

INJECTATE = 0.2 ml
^{22}Na 3 mCi, 1.5 mM
^{3}H-sucrose 15 mCi

Fig. 3. ^{22}Na permeability study following intraventricular or intracarotid treatment with atrial natriuretic factor/hormone (ANF). icv. = intraventricular. + = statistically different from control; $p < 0.05$

a manner very similar to the kidney and other membranes known to regulate fluid and electrolyte permeability (*i.e.* intestine, etc.)[7]. Brain capillaries exhibit anatomical and biochemical features characteristic of secreting epithelia, such as tight junctions, a high mitochondrial content in the endothelial cells, and a similar complement of intracellular enzymes[3].

Brain capillaries are functionally innervated by serotoninergic and adrenergic neurons originating from the brain stem[17]. Nerve fibers containing acetylcholine, vasoactive intestinal peptide, histamine, dopamine, and substance P, etc. have also been found[12, 19, 34, 50, 52]. Electronmicroscopic studies have since confirmed the actual presence of true synapses on a few capillaries[19, 50]. In an immunoelectronhistochemical study we have demonstrated that AVP-containing neural processes in contact with the blood vessels correspond mainly to dendritic structures, as found by Sofroniew *et al.* (1981)[20, 45]. Such release from dendritic processes and somatic sites has been demonstrated[34, 45]. AVP release has also been revealed by morphological studies in the hypothalamic magnocellular areas[23]. Thus, it is possible that, when released, AVP may act via the glial processes or directly on the capillary bed. A few studies have shown the presence of adrenergic receptors acting via the cyclic AMP system on brain capillaries[15, 18]. Owman and Edvinsson (1977) have reviewed known hormonal effects an brain capillaries[34]. Because of the limited part played by capillaries in content of cerebrovascular resistance, and their role in exchange proc-

esses, it is very probable that this capillary innervation is concerned with regulation of the exchange of substances between the blood and the brain.

The functional characteristics of brain capillaries are very similr to those of the membrane which regulate the water and electrolyte permeabilities. Although neither AVP nor angiotensin II affect cyclic AMP levels in isolated brain microvessels, both alter brain water permeability when injected into the cerebral ventricle[37]. Rosenberg *et al.* (1986) even found that centrally administered AVP increased the permeability of ependyma to water[41].

Effects of AVP and ANF on Brain Water and Electrolytes

With these considerations in mind, we have attempted to elucidate the possible involvement of the central AVP system in the development of brain oedema[8–10]. Central administration of arginine vasopressin (125 ng) or DDAVP (0.5 µg), with or without an accompanying water load, brought about a 1–1.3% water accumulation. Changes in the ion contents were inconclusive as to the mechanism of water accumulation. The use of Brattleboro diabetes insipidus rats, known to be unable to synthetize AVP seemed to offer a unique chance for study of the consequence of the lack of this hormone on the development of water balance disturbances after pathological conditions such as subarachnoid haemorrhage (SAH)[9]. Brain water accumulation developed later in the Brattleboro DI animals than in control rats but it then exceeded the level in the controls. The different time-course of brain water accumulation in the Brattleboro DI rats could be ascribed to the following phenomena: early brain water uptake could have been delayed as a consequence of reduction of the capillary permeability increase because of the lack of an acute central vasopressin release after SAH. The 2.6% (Brattleboro) versus 1.4% (control) increase in water content 24 hours after SAH might be explained by the failure of absorption to increase and by the decreased secretion of CSF due to the absence of AVP. This means that AVP may exert opposite actions during the early and later phases of SAH: by enhancing the permeability of the capillaries, the acute, major, short-term increase in the AVP level after SAH may lead to an eventual water uptake, outweighing the opposite action on the choroid plexus and the resorptive apparatus, while the later increased drainage and lowered production of CSF may lead to reduction of water content[9, 28, 46, 47]. In a different model of brain oedema, Reeder *et al.* (1986) likewise confirmed that pharmacological doses of central AVP facilitate the production of vasogenic oedema.

In a recent set of experiments we found that central administration of synthetic rat atrial natriuretic hormone (2 µg) prevented the water accumulation elicited in rat brain by a systemic hypoosmolar fluid load, and led to a significant sodium loss from the nervous tissue[11, 14, 21, 30, 31, 35, 44]. With the modified brain uptake method of Oldendorf, using ^{22}Na and ^{14}C-labelled sucrose, we proved that centrally released atrial natriuretic hormone alters the sodium permeability of the capillary endothelium, which cannot be influenced by the Na-pore inhibitor amiloride or Na-cotransport inhibitor furosemide[1-3].

Conclusion

We are of the view that elucidation of the role of central hormones such as AVP or atrial natriuretic hormone would contribute much towards an understanding of the cellular and biochemical background of brain volume and fluid regulation in various conditions such as raised intracranial pressure. A better understanding of these hormones, their receptors and their pharmacological manipulations have exciting clinical implications.

References

1. Arieff AI, Llach F, Massry SG (1976) Neurological manifestation of hyponatraemia: correlation with brain water and electrolytes. Medicine (Baltimore) 55: 121–140

2. Betz AL (1986) Transport of ions across the blood-brain barrier. Fed Proc 45: 2050–2054

3. Bradbury MWB (1979) The concept of a blood-brain barrier. John Wiley and Sons, Chichester New York Brisbane London

4. Brownfield MS, Kozlowski GP (1977) The hypothalamo-choroidal tract I. Immunohistochemical demonstration of neurophysin pathways to the telencephalic choroid plexus and cerebrospinal fluid. Cell Tissue Res 178: 111–127

5. Buijs RM, Swaab DF, Dogterom J, Van Leeuwen FW (1978) Intra- and extrahypothalamic vasopressin and oxytocin pathways in the rat. Cell Tissue Res 186: 423–433

6. Cserr HF (1974) Relationship between cerebrospinal fluid and interstitial fluid of brain. Fed Proc 33: 2075–2078

7. Cserr HF, Cooper DN, Suri PK, Patlak CS (1981) Efflux of radiolabeled polyethylene glycols and albumin from rat brain. Am J Physiol 240: F 319–F 328

8. Dóczi T, Szerdahelyi P, Gulya K, Kiss J (1982) Brain water accumulation after the central administration of vasopressin. Neurosurgery 11: 402–407

9. Dóczi T, László FA, Szerdahelyi P, Joó F (1984) The role of vasopressin in brain edema formation: further evidence obtained from the Brattleboro diabetes insipidus rats with subarachnoid hemorrhage. Neurosurgery 14: 436–440

10. Dóczi T, Szerdahelyi P, Joó F (1984) 5-hydroxytryptamine, injected intraventricularly, failed to increase brain water content. Neurosurgery 15: 165–169

11. Dóczi T, Joó F, Szerdahelyi P, Bodosi M (1987) Regulation of brain water and electrolyte contents: the possible involvement of central atrial natriuretic factor (ANF). Neurosurgery 21: 454–458

12. Edvinsson L, Copeland JR, Emson PC, McCulloch J, Uddman R (1987) Nerve fibers containing neuropeptide Y in the cerebrovascular bed: Immunocytochemistry, radioimmunoassay, and vasomotor effects. J Cerebr Blood Flow Metab 7: 45–57

13. Fishman RA (1959) Factors influencing the exchange of sodium between plasma and cerebrospinal fluid. J Clin Invest 38: 1698–1708

14. Glembotski C, Wildey GM, Gibson TR (1985) Molecular forms of immunoreactive atrial natriuretic peptide in the rat hypothalamus and atrium. Biochem Biophys Res Com 129: 671–678

15. Harik SI, Sharma VK, Weatherbe JR, Warren RH, Banergee SP (1980) Adrenergic receptors of cerebral microvessels. Europ J Pharmacol 61: 207–208

16. Harik SI (1986) Blood-brain barrier sodium/potassium pump: modulation by central noradrenergic innervation. Proc Nat Acad Sci 83: 4067–4070

17. Hartman BK, Zide S, Udenfriend (1972) The use of dopamine hydroxylase as a marker for the central noradrenergic nervous system in rat brain. Proc Nat Acad Sci 69: 2722–2726

18. Herbst TJ, Raichle ME, Ferrendelli JA (1979) Beta-adrenergic regulation of cAMP concentration in brain microvessels. Science 204: 330–332

19. Itakura T, Yamamoto K, Tobyama M, Shimizu N (1977) Central dual innervation of arterioles and venules in the brain. Stroke 8: 360–365

20. Jójárt I, Joó F, Siklós L, László FA (1984) Immunoelectron-histochemical evidence for innervation of brain microvessels by vasopressin-immunoreactive neurons in the rat. Neurosci Lett 51: 259–264

21. Kawata M, Ueda S, Nakao K, Morri N, Kiso I, Imura H, Sano Y (1985) Immunohistochemical demonstration of alpha-atrial natriuretic polypeptide-containing neurons in the brain brain. Histochemistry 83: 1–3

22. Kimelberg HK, Naunri S, Biddlecome S, Bourke RS (1978) Enzymatic and morphological properties of primary rat brain astrocyte cultures and enzyme development in vivo. Brain Res 153: 55–77

23. Krisch B (1980) Non-granular vasopressin synthesis and transport in early stages of rehydration. Cell Tissue Res 207: 89–107

24. Lindvall M, Edvinsson L, Owman C (1978) Sympathetic nervous control of cerebrospinal fluid production from the choroid plexus. Science 201: 176–178

25. Lindvall M, Owman C (1978) Early development of noradrenaline containing sympathetic nerves in the choroid plexus system of the rabbit. Cell Tissue Res 192: 195–203

26. Lindvall M, Edvinsson L, Owman C (1979) Effect of sympathomimetic drugs and corresponding receptor antagonists on the rate of cerebrospinal fluid production. Exp Neurol 64: 132–145

27. Liszczak TM, Black PMcL, Foley L (1986) Arginine-vasopressin causes morphological changes suggestive of fluid transport in rat choroid plexus epithelium. Cell Tissue Res 246: 378–385

28. Luerssen TG, Robertson GL (1980) Cerebrospinal fluid vasopressin and vasotocin in health and disease. In: Wood JH (ed) Neurobiology of cerebrospinal fluid I. Plenum Press, New York, pp 613–623

29. MacKenzie ET, McCulloch J, Harper MA (1976) Influence of endogeneous norepinephrine on cerebral blood flow and metabolism. Am J Physiol 231: 488–498

30. Morii N, Nako K, Sugawara A, Sakamoto M, Suda M, Shimokura M, Kiso Y, Kihara M, Yamori Y, Imura H (1985) Occurrence of atrial natriuretic polypeptide in brain. Biochem Biophys Res Com 127: 413–419

31. Morii N, Nakao K, Kihara M, Sakamoto M, Sugawara A, Shimokura M, Kiso Y, Yamori Y, Imura H (1986) Effects of water deprivation and sodium load on atrial natriuretic polypeptide in rat brain. Inter-American Society Proc Suppl I Hypertension 8: 161–165

32. Nakamura S, Milhorat TH (1976) Structure and function of the choroid plexus and other sites of CSF formation. Int Rev Cytol 47: 225–288

33. Nathanson JA (1976) Beta-adrenergic sensitive adenylate cyclase in secretory cells of the choroid plexus. Science 204: 843–844

34. Owman C, Edvinsson L (eds) 1977) Neurogenic control of brain circulation. Pergamon ress, Oxford, pp 39–152

35. Palluk R, Gaida W, Hoefke W (1985) Minireview: Atrial natriuretic Factor. Life Sci 36: 1415–1425

36. Peachey LD, Rassmusen H (1961) Structure and function of toad urinary bladder as related to its physiology. J Biophys Biochem Cytol 10: 529–553

37. Raichle ME, Grubb RL (1978) Regulation of brain water permeability by centrally released vasopressin. Brain Res 143: 191–194

38. Raichle ME (1981) Hypothesis: A central neuroendocrine system regulates brain ion homeostasis and volume. In: Martin JB, Reichlin S, Bick KL (eds) Neurosecretion and brain peptides. Raven Press, New York, pp 329–336

39. Rennels ML, Gregory TF, Blaumanis OR, Fujimoto K, Grady PA (1985) Evidence for a "paravascular" fluid ciruclation in the mammalian central nervous system, provided by the rapid distribution of tracer protein throughout the brain from subarachnoid space. Brain Res 326: 47–63

40. Rodriguez EM (1976) The cerebrospinal fluid as a pathway in neuroendocrine integration. J Endocrinol 71: 407–443

41. Rosenberg AG, Kyner WT, Fenstermacher JD, Patlak CS (1986) Effect of vasopressin on ependymal and capillary permeability to tritiated water in cat. Am J Physiol 251: F 485–F 489

42. Rudman D, Hollins BM, Lewis NC, Scott JW (1977) Effects of hormones on cAMP in choroid plexus. Am J Physiol 323: E 353–E 357

43. Schultz WJ, Brownfield MS, Kozlowski GP (1977) The hypothalamo-choroidal tract II. Ultrastructural response of the choroid plexus to vasopressin. Cell Tissue Res 178: 129–141

44. Skofitch G, Jacobowitz DM, Eskjay RL, Zamir N (1985) Distribution of atrial natriuretic factor-like immunoreactive neurons in the rat brain. Neuroscience 16: 917–948

45. Sofroniew MV, Glasman W (1981) Golgi like immunoperoxidase staining of hypothalamic magnocellular neurons that contain vasopressin, oxytocin or neurophysin in the rat. Neuroscience 6: 619–643

46. Sorensen PS, Gjerris F, Hammer M (1982) Cerebrospinal fluid vasopressin in benign intracranial hypertension. Neurology 32: 1255–1259

47. Sorensen PS, Gjerris F, Hammer M (1984) Cerebrospinal fluid vasopressin and increased intracranial hypertension. Ann Neurol 15: 435–440

48. Sorensen PS, Vilhardt H, Gjerris F, Warberg J (1984) Impermeability of the blood-cerebrospinal fluid barriers to l-deamino-8-D-arginine-vasopressin (DDAVP) in patients with acquired communication hydrocephalus. Eur J Clin Invest 14: 435–439

49. Steardo L, Nathanson JA (1987) Brain barrier tissues: end organs for atriopeptins. Science 235: 470–473

50. Swanson LW, Connelly MA, Hartman BK (1977) Ultrastructural evidence for central monoaminergic innervation of blood vessels in the paraventricular nucleus of the hypothalamus. Brain Res 136: 166–173

51. Wang BC, Share L, Goetz KI (1985) Factors influencing the secretion of vasopressin into the cerebrospinal fluid. Fed Proc 44: 72–77

52. Wood JH (1982) Neuroendocrinology of cerebrospinal fluid: Peptides, steroids, and other hormones. Neurosurgery 11: 293–305

Correspondence: T. Dóczi, M.D., Department of Neurosurgery, University Medical School, Pf: 464, H-6701 Szeged, Hungary.

Acta Neurochirurgica, Suppl. 47, 127–128 (1990)

The Hypothalamus: New Ideas on an Old Structure

M. Brock

Neurochirurgische Klinik, Klinikum Steglitz der Freien Universität Berlin

The present seminar has brought up many of the fascinating ideas presently accepted concerning the functions of the hypothalamus. Summarizing them is an extremely difficult task. At the same time it constitutes a challenge, since the subject dealt with is of great practical importance, not only due to its interdisciplinary range but also to its integrative potential much in the same way as the hypothalamus itself appears to be one of the main integrative centres of the brain (if not the major one). It is certain that when Sherrington wrote his *Integrative Action of the Nervous System*[1] he was far from thinking in the terms of today, but already the classical work of Papez[2] represented a preview of modern concepts. The conceptual shifting from a topographic mapping of hypothalamic "centres" towards a more dynamic approach thanks to biochemical studies has, no doubt, led to a new era in neuroendocrinology. It is now known that the afferents to the hypothalamus are not necessarily neuronal but may also be humoural (both CSF- and blood-borne). Further, it has become apparent that hypothalamic cell groups concentrate medially, perhaps seeking for proximity to the CSF spaces, perhaps because this is the deepest and most protected part of the brain, while the lateral hypothalamus is rich in fibers and appears to constitute part of a wide communication system. Efferents from this system reach not only centres in the medulla oblongata, such as the locus coeruleus, but may extend even to the lumbosacral levels in the intermediate grey column. These efferents emanate from the dorsal and from the lateral parvocellular districts of the paraventricular nuclei, and many of them use oxytocin as transmitter substance.

Amazingly enough, many hypothalamic neurons may produce several substances. Thus, one and the same neuron, for instance in the arcuate nucleus, may secrete dopamine, growth releasing factor, neurotensin, galanin and, perhaps, GABA. The question remains open as to when—and why—the one or the other substance is produced. The same situation appears to apply to the cells of the parvocellular part of the paraventricular, which are involved in the control of ACTH by the anterior pituitary, and which produce encephalin, neurotensin, vasoactive intestinal peptide (VIP) and corticotrophic releasing factor (CRF).

The open questions related to the *pattern of vascularization* and the direction of blood flow in the hypothalamic-pituitary system make the situation even more complex. Since the anterior pituitary does not have its own arterial supply, and since there is a large species variation in the vascular anatomy of this area, it appears difficult to ascertain the role of blood flow in this neuroendocrine system. While this blood flow, reaching levels of 500 to 700 ml/100 g/min, is 10 times higher than cerebral blood flow and is paralleled only by choroid plexus blood flow, it appears not to display the 3 main characteristics of cerebral blood flow: functional regulation, CO_2-dependence, and autoregulation. Further, intrasellar pressure gradients in cases of intrasellar expanding processes might cause pituitary stalk compression and impair portal blood flow both by direct vessel distortion and by reduction of local perfusion pressure.

Another important fact is that *hypothalamic peptides*—both releasing and inhibitory—control or regulate pituitary function. There is at least one hypothalamic peptide for each pituitary hormone, but some of these peptides are not restricted to influencing the secretion of one pituitary hormone only. Most of these hypothalamic hormones are produced as prohormones within hypothalamic neurons, and undergo a series of changes on their way to the pituitary along the hypothalamo-pituitary axons. These changes include, for instance in the case of arginine-vasopressin, translation, packaging, and glycosylation. The factors which switch on the hypothalamo-pituitary axis include virtually all biological phenomena. Sensor sites may be located at very distant areas in the body, and information coming from them may, apparently, undergo modifcation and/or modulation on the way to higher centres.

Since neurosecretion is also accompanied by electrical discharges of the involved neurons, rapid integrative action is possible as well as the induction of fast changes at distant secretory and non-secretory cell groups. In this context, hypothalamo-cortical and cortico-hypothalamic relationships still deserve further studies. It is interesting that, at least under pathological conditions, there may be a dissociation of previously coupled hormonal functions. An example is the isolated abolition of the vasopressor action of vasopressin with preserved osmoregulation in patients with Multiple System Atrophy (MSA), and the opposite functional dissociation in other circumstances.

Mention has been made in the foregoing pages of the ability of some neuroendocrine cells to establish new, reversible, function-dependent connections with other cells. This is the case, for instance, of the oxytocinergic neurons in lactating rats. This transient functional *flexibility* widens the original concept of *plasticity* of biological systems. This plasticity is favoured by the diffuse nature of the neuroendocrine system, with peripheral components extending as far as the pulmonary and gastrointestinal districts. Also, little is known as yet concerning the action of neuropeptides on glial cells. These cells may well become a new and bright nebulosa in the neuroendocrinological firmament. At present they still are, to a large extent, a black hole.

Both plasticity and flexibility, but also normal function, are not possible without timing. Timing means rhythmicity. Rhythmicity requires clocks and oscillators. Recent studies indicate that there is not one major biological clock in the hypothalamus but that several oscillators may react to different stimuli, or may act in an overlapping manner to a same stimulus—another example of plasticity (in the broad sense of the word). The close interrelationship between circadian and ul-

tradian rhythms demonstrates that, in addition to the "milieu interieur", extracorporeal influences, and this obviously includes extraterrestrial forces, act on and through the hypothalamus. Thus, although the hypothalamus may not be the *seat* of the soul, as stated by Carmel, it certainly is the *door* to it.

The major problem still afflicting those interested in treating disorders of the hypothalamo-pituitary system is the large gap between the animal models and the clinical situation. This gap is certainly not restricted to, but particularly distressing in, the field of neuroendocrinology. In man, lesions of this system are not restricted to individual nuclei, are progressive, are of heterogeneous location and are of varying rate of propagation. Removal of the perturbing agent, for example by surgery, may even increase a pre-existing functional deficit. From the clinical point of view, it appears that basic neuroendocrinological research has contributed much more to diagnosis or to the control of therapy than to therapy itself (an exception being bromocryptine). We have seen that surgery, also curative quoad vitam, not uncommonly is deleterious quoad functionem. This is the case even in benign, congenital lesions, such as the craniopharyngioma. This is all the more true in other tumours affecting the hypothalamus. This is even the case in subarachnoid hemorrhage, both by direct mechanical influence of a dilated 3rd ventricle on the hypothalamus and by changes of blood volume, blood viscosity and cerebral blood flow.

If it is true that man was made by God, it is also true that it was in creating the hypothalamus that the Allmighty invested the largest amount of inspiration. There is, still, a long way to go in observing and understanding this inspiration. This has become clear from the scientific contributions to this seminar.

References

1. Sherrington CS (1926) The integrative action of the nervous system. Yale University Press, New Haven
2. Papez JW (1937) A proposed mechanism of emotion. Arch Neurol Psychiatry 38: 725–743

Correspondence: Prof. Dr. med. M. Brock, Neurochirurgische Klinik, Klinikum Steglitz der Freien Universität Berlin, D-1000 Berlin 45.

New by Springer-Verlag

Vinko V. Dolenc

Anatomy and Surgery of the Cavernous Sinus

Foreword by Mahmut G. Yaşargil

1989. 182 figures (approx. 450
single illustrations).
XII, 344 pages.
Cloth DM 280,–, öS 1960,–
ISBN 3-211-82155-4

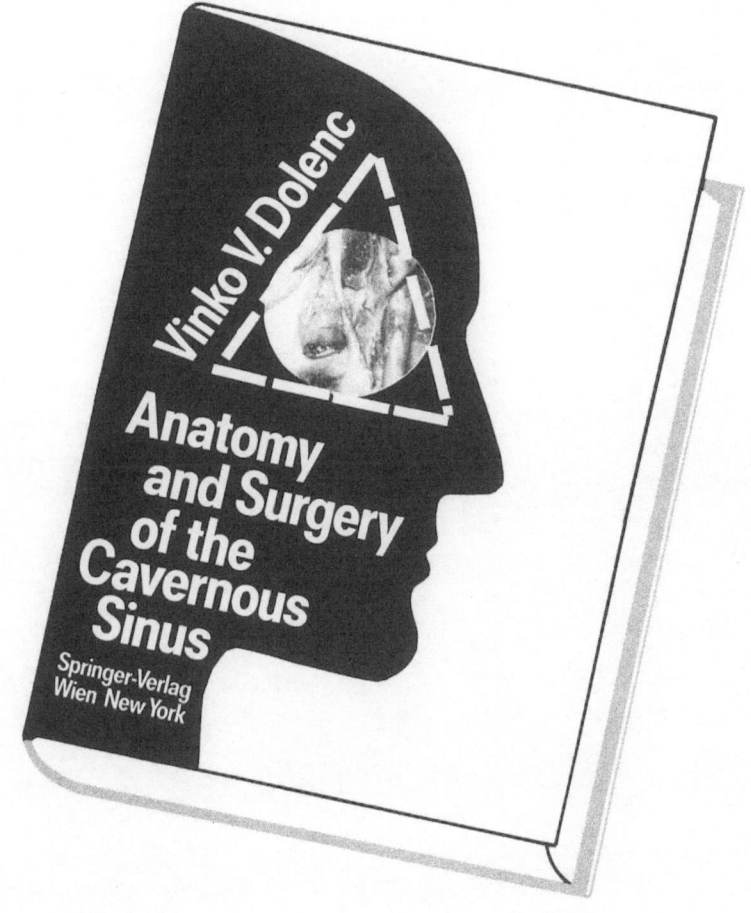

This atlas discusses the surgical anatomy of the parasellar and neighbouring regions in the midline and in the posterior cranial fossa. Its aim is to demonstrate the importance of normal anatomical structures and interrelationships in the management (treatment) of pathology in this area. Therefore, Chapter 1 is devoted exclusively to the anatomy of the cavernous sinus and neighbouring regions. Anatomical dissections of fresh specimens were prepared after injection of the arterial and venous systems. The surface of the cavernous sinus and neighbouring regions were, with reference to standard anatomical relationships, divided into ten triangular "windows" through which it is possible to enter the cavernous sinus and to study the normal relationships between individual structures. The most important feature of this book is the parallel drawn between surgical anatomy as demonstrated on normal fresh cadaver specimens and in vivo surgical procedures for tumorous and vascular lesions, which are only possible if the anatomical relationships on the surface of the cavernous sinus and its walls, the architecture of its interior and the previously determined sequence of operative approaches are taken into account. It is clearly demonstrated in detail that there is no one single approach to vascular and tumour pathology of the cavernous sinus. On the contrary, there are a number of possible approaches which are suitable for various pathological conditions depending on their locations. It is intended that everyone who studies this book will be able to distinguish important anatomical landmarks and orientation on the surface of and inside the cavernous sinus, which will aid in both the surgical management of patients and in clear and precise communication.

This book is the first to deal with the anatomy of the cavernous sinus in a practical surgical manner and which applies the anatomy to concrete examples. 126 colour photos, 134 drawings, and 49 tableaux of angiograms, CTs and MRI images provide a great degree of clarity.

Springer-Verlag Wien New York
Moelkerbastei 5, P.O. Box 367, A-1011 Wien
Heidelberger Platz 3, D-1000 Berlin 33
175 Fifth Avenue, New York, NY 10010, USA
37-3, Hongo 3-chome, Bunkyo-ku, Tokyo 113, Japan

New by Springer-Verlag

Rezio R. Renella

Microsurgery of the Temporo-Medial Region

1989. 50 partly colored
figures (123 single illustrations).
Approx. 220 pages.
Cloth DM 158,–, öS 1110,–
ISBN 3-211-82144-9

The differentiation of the temporal lobe into a lateral neocortical and a medial allocortical region is supported by developmental, anatomical and clinical evidence. Although this view of a dual temporal lobe is generally accepted by neurosurgeons dealing with functional surgery, it still receives little attention by those approaching structural abnormalities located or extending into the medio-basal region. Consequently, the characterization of the temporo-medial area as a distinct surgical region is still lacking. The major object of this study is to analyse the medial part of the temporal lobe as a distinct surgical region and to integrate the microsurgical and physiological aspects into a concept applicable to the several types of temporo-medial lesion. The study includes five sections. The first section is devoted to the morphological aspects. The second and the third sections present a simplified clinical approach to temporo-medial lesions and analyse the ancillary investigations which are indispensable for characterizing their structural and functional features. The fourth section deals with the surgical aspects of temporo-medial lesions, and especially with the selection of the optimal approach having regard to the location of a given process, and to the extent of the functional changes. The last section is devoted to commentaries concerning the neuropathological aspects and the outcome of surgery in the temporo-medial region.